Principles of Web Design

By Brian D. Miller

Foreword by Jason Ackerman

01 | Plan   **02** | Design   **03** | Optimize

Principles of Web Design

By Brian D. Miller

Foreword by Jason Ackerman

ALLWORTH PRESS
NEW YORK

Praise for Previous Editions

One of the Best Books Available on the Web Design Process
Brian clearly put a lot of thought and careful consideration into the structure, content and flow of the book. It's well done. I've been building websites for 4 or 5 years now and this is the first book I've found that does a good job of walking the reader through the entire thought process of planning and creating a structured strategy for designing websites."
– Published on Amazon.com by Steffan Antonas

Web Design Demystified
'Above the Fold' provides everything that you need to build a strong foundation for a successful Website."
– Published on Amazon.com by Ted A. Dobbs

I'm Glad I Started Here
Brian Miller gives a sensible and well thought out approach to web design. As a person who is interested in getting into this field I found this book a great place to begin my journey. He links all the steps to be successful at web design. The talent and skills must be your own, but if you're looking for a tool to focus existing talent into a career in web design, this book is an excellent place to begin."
– Published on Amazon.com by JStein

Fantastic Guide

Above the Fold provides you with the fundamentals required for a successful digital experience. Anyone who is looking to promote themselves or business using digital media marketing tools should own a copy of this book. As Brian Miller thoroughly demonstrates, principals of design very much apply to each of your digital properties. Thank you, Brian, for an extremely comprehensive guide of an ever changing landscape. While technology will continue to change, principals of effective design do not."
– Published on Amazon.com by B. Crosland

Great for Budding Devs

If you're into the web and building sites then this, IMHO, is a must read. Brian Miller took his time and laid out the fundamentals of web development and design here in a concise and accessible manner. You don't need a lot of technical acumen or programming prowess to get what he's writing about and you'll come away with a much greater understanding of the how's and why's of dev/design."
– Published on Amazon.com by Frater Zion

A Useful and Easy Guide

Miller has written here the perfect handbook for a student looking to learn more about web design or a small business owner trying to spice their site up and increase traffic. It's all here, discussions about the web itself, about how people use it and read different sorts of pages, discussions about the usability of a site and how that affects its design and vice-versa, a checklist of things to do before your site goes live and of things to have as it grows and thrives, everything you could need to know about web design."
– Published on Amazon.com by A Student of eCommerce

A Must Have for Web Designers

Brian Miller does a great job of introducing people to the Art & Science of Website Design. This book is an excellent introduction and overview of Website Design, but is also packed with great information for old-timers as well."
– Published on Amazon.com by Hugh

Allworth Press books may be purchased in bulk at special discounts for sales promotion, corporate gifts, fund-raising, or educational purposes. Special editions can also be created to specifications. For details, contact the Special Sales Department, Allworth Press, 307 West 36th Street, 11th Floor, New York, NY 10018 or info@skyhorsepublishing.com.

26 25 24 23 22   5 4 3 2 1

Published by Allworth Press, an imprint of Skyhorse Publishing, Inc. 307 West 36th Street, 11th Floor, New York, NY 10018. Allworth Press® is a registered trademark of Skyhorse Publishing, Inc.®, a Delaware corporation.

www.allworth.com

Designed and art directed
by Brian D. Miller
WiltonCreative.com

Every reasonable attempt has been made to identify owners of copyright. Errors or omissions will be corrected in subsequent editions.

Cover design by Brian D. Miller
Interior design by Brian D. Miller

Library of Congress Cataloging-in-Publication Data is available on file.

Print ISBN: 978-1-62153-787-8
eBook ISBN: 978-1-62153-788-5

Printed in China

For my two beautiful girls, Sarah and Rachel,
this book, and everything I do, is for you.

Acknowledgments

Special thanks to my college professor and very good friend for over 30 years, Alex White. *Here's to the next 30, Axe.*

Thanks to the guy who breathes life into our work, Jason Ackerman. The future is Bright!

To my friend and copywriting partner, Greg Voornas (Voornas.com). You taught me what it means to be creative and how to live your blood type—*be positive*!

With great appreciation for the talented team at Axioned (Axioned.com), including Libby and Dave, whose partnership I value greatly.

Thanks to Dan and Kelly at LS Media (LSMedia.com). I am continually inspired by your drive and talent.

To my clients. My goal is always to serve you with expertise, patience and value. Thank you for trusting your brands with me and my team.

And thank you to my family, Bridgette, Sarah, and Rachel. *I love you, always.*

Table of Contents

Foreword

It still happens, no matter how hard we try. In our agency, we get clients who look at the proposal, see the line item marked "Design", and question why we can't just "design it as we build". And there was a time we would just go ahead and do it. I used to consider myself a "designer/developer," the rare unicorn that could do both—a solid eye for design and all the sensitivity of a real artist, combined with the technical prowess and engineering mindset of a real nerd. I was the perfect web guy.

Brian Miller and his design team cured me of that. It was from them that I really learned what design was about in the context of web technology. I learned how much really goes into a quality website design. I learned that it's as much about what is NOT on the screen as what IS on the screen. I learned that everything we were looking at on the first PSD was there for a reason—everything had a purpose. And it wasn't just a purpose, it was a defendable purpose. Every font-size, every margin-left, every padding-top, line-height, letter-spacing, height and width . . . each one was carefully considered and important.

I first met Brian when my now-defunct firm was contracted to build the front end for a website he was designing. Having now worked with Brian for nearly a decade, for a time as his employee and now again as colleagues with my own new and different agency, I find it hard to work with a lot of designers. It's not because they aren't talented, it's because very few put the kind of thoughtfulness and consideration into the WHY of their design. That's what makes this book so valuable of a resource . . . it doesn't try to teach you how to be a talented designer. That's on you. This teaches the why, and intuitively links it to the how.

As a developer, it took quite some time for me to understand the true partnership design and development can have in a project. So often that relationship is confrontational—designers making things that developers have to figure out how to make work, and constantly pushing and pulling between the two until you end up with a final product that is the result of compromise, and it shows. But it doesn't have to be that way. Design that takes into account the elastic medium of the browser, as well as the unpredictable whims of the user, is not only a better end product, but also a joy to build.

By my count, I've worked on close to 100 sites with Brian, each one made better by the principles found in this book. I hope you find it as useful as I have.

Jason Ackerman
Founder, Overtime Agency
www.overtimeagency.com

Introduction
to **Web Design**

There's an old legend in the world of football that says Vince Lombardi, head coach of the Green Bay Packers, started every season with a speech to his players about the game of football. He began the lecture by holding up a football and saying, "Gentlemen, this is a football." He proceeded to describe its size and shape, and talk about how it could be thrown, kicked, and carried. Then he'd point down at the field and say, "This is a football field." He'd walk around, describing the dimensions, the shape, the rules, and how the game was played.

This Is the Internet

The message from the two-time Super Bowl-winning coach was simple: to truly be effective at anything, one can never forget the basics. This simple demonstration stripped away the complexities of the game and reduced it to its essence. In doing this, Lombardi refocused his team's attention on what was truly important about succeeding at the game of football.

Taking a cue from Vince Lombardi, I'd like to conduct a similar exercise for you: Turn on a web-enabled device (PC, laptop, tablet, mobile phone, etc.), open the web browser of your choice (Safari, Chrome, Firefox, Microsof Edge, etc.), type in the address of your favorite website, and behold—this is the internet. The internet is a series of interconnected computers, called servers, that enables companies, brands, organizations, governments, religious groups, and individuals to share information on a worldwide scale in real time. The World Wide Web or web, for short, is actually only a portion of the internet, which also includes all aspects of computer-to-computer communication like email, messaging, and file serving, just to name a few.

When an internet user types the address of a website into his or her web browser, the device transmits a signal to a server, and the server responds by sending bits of information back to the computer. This information includes images, raw content, and instructions for the computer to reassemble the layout, called markup (the M in HTML). The computer then takes that information and configures the files based on two things: the markup and styles that came from the designer/developer, and the preferences and limitations of the web browser and device itself. When a device reassembles a web page that it has received from a server, the following factors influence exactly how that page appears on the screen.

DEVICE

The type of device and version of the operating system (OS) the audience is using to browse a site can have an effect on how a site is seen. The number of operating systems has increased over recent years. Instead of focusing on Apple versus Microsoft, designers and developers now have mobile platforms to contend with—iOS (Apple), Windows Mobile, Android (Google), and to some extent Blackberry. A primary difference between operating systems is how typography is handled, including the fonts that are available natively and how smoothly the fonts are rendered. Chapter 6 takes an in-depth look at typography.

SCREEN RESOLUTION

Not to be confused with the screen size in inches, resolution is the dimension in pixels measured horizontally and vertically on a screen. Most desktop monitors range from 800 pixels wide by 600 pixels high to 1024 pixels wide by 768 pixels high—and high resolution, or Retina Displays, can reach as much as 3072 x 1920. Tablets have similar resolutions, while mobile devices can be as little as 320 pixels wide. Because of this dilemma of differing screen resolutions, designers and developers created the idea of responsive design. Discussed further in chapter 3, responsive design displays different layouts for a single web page in response to the screen resolution, making it possible to maximize legibility and usability regardless of the size of the screen on which the content is displayed.

WEB BROWSER

The primary web browsers used today are Safari, Chrome, Firefox, and Microsoft Edge, both desktop and mobile versions. A web browser is an application whose function is to receive layout and styling information from a host and display that information on screen. Because these are different applications developed by different companies, they all interpret this information slightly differently. Added to this, the language that makes up web styling—cascading style sheets, or CSS—is always evolving; therefore, web browsers are constantly updating to keep up with the latest styling attributes.

CONNECTION SPEED

The connection speed is the speed with which a computer or device can connect to the internet and download the assets required to build a page. This has been an on-again, off-again issue through the years. The first computers to connect to the internet did so with modems that used phone lines, which were very slow, causing the need for "lightweight" pages—pages created mostly of text and color, with few images. Then came DSL and cable modems, making high-speed internet possible, and web page design evolved to include large amounts of imagery. Enter the cell phone, and people began browsing the web with slower connection speeds, until Wi-Fi and high-speed mobile connections evolved. While the connection speed of a user browsing a site won't have a direct effect on how a site looks, it will definitely have an effect on the person's experience of the site.

Designing for the Web

To complicate matters, beyond these inherent system-based influences, individual user preferences also can affect the way a site looks. In this image we see the "Content" preferences in the Firefox web browser. These controls allow a savvy web user to change the fonts, the minimum size for type (this is an accessibility feature for users with impaired vision), the colors used for links, and even whether links are underlined. In some cases, these user preferences can even override the design decisions a designer has made for a page.

It is this aspect of disassembling a design and allowing the user to reassemble it under a varying set of circumstances that makes web design a unique and challenging form of design. These unique factors create added limitation considerations, and new possibilities, for the designer. Dealing with these factors and the potential issues they can cause in the clear communication of a message or a brand image requires a specific process.

The influence of the web browser on web design can be seen clearly in the following timeline. As the browser evolved, so did the sophistication of the design treatments for web pages. Also evident on the following timeline is the uniquely web idea of "publish, then polish." For many web-based organizations, like the ones in the timeline, getting something online is more important than getting the perfect thing online. This can be very counterintuitive for print designers, who are used to meticulous perfection prior to any public consumption.

This screenshot of the preferences panel in Firefox shows how users can change how specific characteristics of web design appear on their screen.

Brief History of Web Design

Web 1.0 (1993 – 1997)

1993

Mosaic, the first consumer Web-browsing application, is released

1994

Yahoo.com launches

WC3 is formed to standardize HTML

Netscape Navigator Web browser is released

1995

Amazon.com launches

NYTimes.com launches

CraigsList.org launches

Microsoft releases Internet Explorer versions 1 (August) and 2 (November)

1996

CompuServe changes its name to Lycos.com

Cascading style sheets (CSS) introduced

Microsoft releases Internet Explorer version 4

Weather.com launches

1997

DrudgeReport.com launches

Ebay.com launches

Netscape Communicator replaces Netscape Navigator

A Brief History of Web Design

Web 1.0 (1998 – 2007)

1998

Google.com, founded by Larry Page and Sergey Brin, launches

1999

Napster.com, a peer-to-peer file sharing Website, launches

Microsoft releases Internet Explorer version 5, which allowed users to save web pages for the first time

2000

Craigslist.org expands beyond San Francisco (originally launched in 1995)

Google Adwords launches

Netscape version 6 is released

2001

Wikipedia.org launches

Microsoft releases Internet Explorer version 6, which included support for CSS

2002

Friendster.com launches

Netscape version 7 is released

2003

MySpace.com launches

WordPress blogging software is introduced.

Apple releases the Safari web browser

2004

Facebook.com launches

Flickr.com launches

Mozilla Firefox web browser is released, which utilizes the Gecko layout engine to display web pages

2005

YouTube launches

Reddit.com launches

2006

Twitter launches

Microsoft releases Internet Explorer version 7, which introduced tabbed browsing and a content feed reader

Mozilla Firefox version 2 is released with tabbed browsing

2007

Apple introduces the iPhone and mobile apps

Netscape Navigator version 9 is released

Candidate Websites and social media play a pivotal role in the U.S. elections

Mozilla Firefox version 3 is released

Microsoft launches Bing.com to compete with Google

Microsoft releases Internet Explorer version 8 with improved support for Ajax, CSS, and RSS

Twitter.com is used to organize and mobilize relief efforts in Haiti following the devastating earthquake.

Mozilla Firefox version 3.6 is released

Pinterest.com launches

SnapChat.com launches

Apple Siri is introduced

Vine.com launches

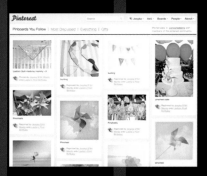

2013

GIFY.com
launches

2014

Amazon Echo—
Alexa — launches

Healthcare.gov
launches

Twitch.com
launches

2015

Google Photos
launches

YouTube Kids
launches

2016

TikTok.com
launches

2017

Facebook Watch
launches

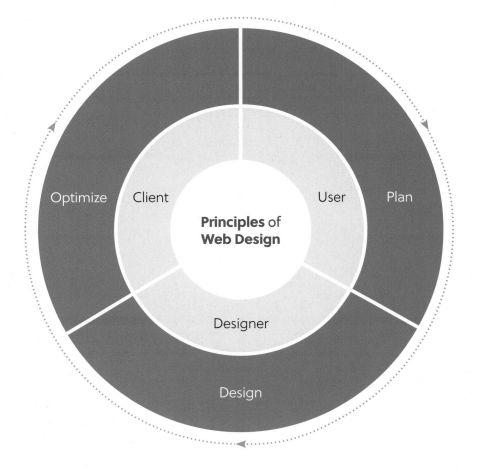

Principles of Web Design *focuses on the three phases of a web project—planning, designing, and optimizing—with each phase aligning with the constituents of a Website: the user, the designer, and the client.*

Plan

The first section of *Principles of Web Design* explores the steps that are required to plan out a website. Planning is not unique to web design, of course, but there are some unique considerations a web designer must be aware of in order to avoid common pitfalls. Even the simplest of websites can be defined as an application with a unique set of utilities that need to be manipulated by the user. Identifying the requirements of this application, including the goals of the client, is a great first step for any web project. The resulting requirements document can be referred to throughout the entire project to create success factors to be used to evaluate the project in the end.

Another benefit of this planning stage is that it helps designers break up large tasks into manageable smaller tasks. Mapping out the relationships between large amounts of complex information or detailing the flow of a particular user task are examples of things that should be addressed prior to beginning the design phase in order to make sure they're getting the attention they require.

In addition to having a plan, web designers need to have a contingency plan—a backup plan that allows for user variables.

The collection of these plans is called User Experience Design, or UX. Designing the experience that's right for the target customer (in addition to what we traditionally think of as graphic design—styling, typography, and imagery) is critical to being a successful web designer. It's the criteria by which each of the samples shown in this book has been judged. They go beyond looking good: they look good, they work well, and in many cases they add an element of delight to the experience. It's also the criteria that the web-browsing population uses to determine how successful a website will be.

Take Twitter.com, for example—a website unlikely to win a traditional design award, yet undeniably and profoundly popular. Twitter's popularity is largely due to two main things: It's a simple idea, telling your followers what you're up to; and it is executed simply, with an emphasis on user interaction. It is a utility that lets users have enough control over the experience to make them feel as if they're expressing themselves, but not so much control that the experience becomes overwhelming or intimidating. This is all a direct result of excellent planning and user experience design — if perhaps not graphic design.

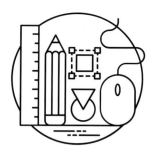

Design

The second section of *Principles of Web Design* looks at the specific attributes of web design and layout. To explore the subject of design for any medium, it's important to define the term *design*. At its most basic, design is a plan. Things that are said to have happened "by design" are said to have happened not by accident.

A finished design is simply the result of a series of decisions made by a designer to express a specific brand image and communicate a message. Each decision a designer makes leaves him or her open for subjective criticism, and therefore, many designers find it helpful if their decision set is limited in some way—by brand guidelines, client requests, or self-imposed limits. Limiting subjective decisions and being creative within those limitations is the essence of what all designers do.

The web as a design medium comes with several built-in design decision limitations—from color accuracy to typographic control to page size. Successful web designers embrace these limitations and find ways to be creative within them, instead of trying to circumvent them. Section II of *Above the Fold* explores the aspects of graphic design (space use, typography, imagery) in the context of the limitations and opportunities that web design offers.

Design is about having a plan. Web design is about having a **backup plan.**

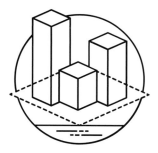

Optimize

The final step in the web design process, as well as the last section of *Principles of Web Design*, is the analysis phase. Analysis can begin with the product itself—the website that was created in the design phase. This testing, or beta, stage can help uncover issues with the digital product prior to launching the site to the public.

Once the site is launched, how will your users find it? Section III looks at two ways of attracting visitors: search engine optimization (SEO) and web marketing. SEO is not a sexy topic. It involves research, copywriting, and networking. But it is paramount to the success of a site. Marketing is very sexy—social, viral, guerrilla. It is these concepts of SEO and marketing that bring users to a site and ultimately lead to its business success.

Finally, web design offers an unprecedented opportunity to analyze and adjust a design based on detailed, real-time information. Improvements to the design or usability of a site can be done on the fly with no limits to the number of changes that can be made. Analytic software, such as Google Analytics, provides countless pieces of data that help a designer understand the habits of the users of a site.

Each of the topics in *Principles of Web Design*, from planning and design to marketing and analysis, can be researched in much greater depth than what is presented here. It is also equally important to take in the breadth of these principles. They are interconnected; too great a focus on one area over another will result in a less than successful product. Planning, designing, analyzing, and back again to planning is the complete and necessary cycle for successful and long-lasting web strategies.

Section I

Plan

1. Website Planning
2. Elements of Usability
3. Space, Grids, and Responsive Design

CHAPTER 1

Website Planning

User-focused design, or design that puts the user ahead of stylistic design treatments or gratuitous use of technology, must start with a plan. The objective of this plan is to align the client's business goals with the needs and desires of the target user group. A plan can also help map out a "big picture" view of the project, giving all members of the team perspective, clarity, and a common goal. An effective plan helps remove subjectivity from the creative process and gives a framework for decision-making.

Project Planning

Creating a Website project plan is a multi-part, multi-disciplinary process. The phases of this process can include research and discovery, content inventory, site mapping, wireframing, usability mapping, prototyping, and design concepting, all of which are discussed in this chapter. Depending on the size of the project, this phase can take a week to several months to establish the documents needed to effectively move forward with the design phase.

There are many benefits to developing an effective site plan. The client should reap long-term benefits, from a reduction in the development cost normally associated with inflexible or flawed systems, to decreased training costs. These benefits help clients make the most of their Website and achieve the highest return on their investment (ROI).

Plans also help the design team define the parameters of a project for estimating purposes. Once a plan is in place, the designer or project team should have a clear picture of the scope of work (SOW) for the project. The team can then estimate and assign time to each task or phase of the project. If along the way the client has revisions or changes direction, the designer or project team can refer back to the approved plan and determine whether the project needs to be re-estimated or if the alterations are within the original scope of work.

Ultimately, however, site planning should be about the user. The goal of a well-conceived site plan is to increase a user's satisfaction with a site by organizing information and optimizing the critical tasks on the site. The measure of the ease of use for a site is called usability and is discussed in the next chapter. What follows are the basic steps involved in the website planning stage.

Research & Discovery

The process of developing a plan usually starts with research into the client's goals for the site and an analysis of the landscape in which a site will exist. A briefing meeting is an interview with the client to better understand the purpose behind the project. This can be conducted by a designer or an account executive (also called a client manager), whose job is to manage the client relationship. A SWOT (strengths, weaknesses, opportunities, and threats) analysis can be very helpful in pinpointing the internal and external factors that will influence the project. A SWOT analysis categorizes the internal, external, positive, and negative factors that can influence the effectiveness of a site.

To gain a deeper understanding of the landscape, it's often necessary to conduct a competitive analysis and customer interviews. A competitive analysis results in noting what the competition does well, as well as where they fall short. This can help identify gaps in the market that the client can take advantage of. Customer interviews are helpful for identifying the current perception of the client organization or the general feeling of the current market.

The result of a client briefing and customer interviews is a project or creative brief. A creative brief outlines the goals for a project, the special considerations the team must take in order to complete the project effectively, as well as a schedule of milestone events. A brief is usually reviewed by the team and the client and signed off by both, forming the directional foundation for the project.

To the right is an example of a SWOT analysis. The process of developing a SWOT chart can help uncover key pieces of information that help shape the usability and concept of a website. Strengths and weaknesses are internal factors, while opportunities and threats are external factors that a client has little control over.

A **SWOT analysis** categorizes the internal, external, positive, and negative factors that can influence the effectiveness of a site.

Strengths
Internal/Positive

Recognized brand

Impressive product line

Weaknesses
Internal/Negative

Understaffed

Lack of experience

Opportunities
External/Positive

Expanding customer base

Growing industry

Threats
External/Negative

Strong competition

Economic factors

Requirements Documentation

A great way to organize the client's needs and create a list of success factors is with a requirements document. A requirements document is usually a spreadsheet that contains a list of non-subjective "must-haves" for each page or section of a site, as well as global must-haves for the whole site. An example of a requirement is "the site shall have commerce functionality," or "the main navigation shall include a link to the shopping cart feature." These requirements help set a framework for the rest of the planning stage and they can be referred to throughout the project to ensure the success of the project.

(Opposite) This creative brief template helps clients synthesize the goals of a project in a way that can help remove a lot of the subjectivity that comes with creative projects.

(Below) A Gantt chart shows the timing of the tasks involved in a project in relationship to one another, helping the team see the big picture.

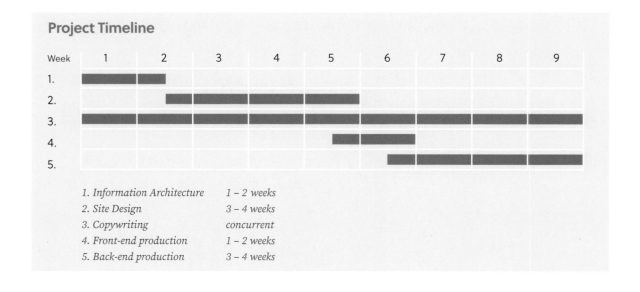

Project Timeline

Week	1	2	3	4	5	6	7	8	9
1.	■	■							
2.			■	■	■				
3.	■	■	■	■	■	■	■	■	■
4.					■	■			
5.						■	■	■	■

1. *Information Architecture*   *1 – 2 weeks*
2. *Site Design*   *3 – 4 weeks*
3. *Copywriting*   *concurrent*
4. *Front-end production*   *1 – 2 weeks*
5. *Back-end production*   *3 – 4 weeks*

Creative Brief

Project name:

Date:

Prepared by:

Submitted to:

Project overview:

Background information:

Target user insight information:

Brand attributes, promise, and mission:

Competitive landscape:

This fact-based portion of the brief should be concise and only include information pertaining to the desired outcome of this specific project

Business objectives—success criteria:

Testing requirements—measurement of success:

Creative strategies:

The business objective should identify a single testing metric that drives the creative strategy and the decision-making process for the project

Functionality and technical specifications:

Contribution and approval process:

Timelines:

Budget:

Defining the number of rounds of revisions and identifying a single point of contact (client and creative) will cause clients to focus their comments and streamline the process

Asset Inventory

26

A website design project can often be overwhelming at the beginning. There are many considerations to be made and items to be collected before designing can begin. Project assets like client logos, copywriting, imagery, and code libraries must all be identified and located. This process begins with an inventory of all the assets needed for a project—in other words, what are the elements of a site that the team will need to complete the project? This information can be collected in a spreadsheet, drawn out on a whiteboard, or sorted on index cards—whatever will produce the most comprehensive results. This process can be done by the creative team in parallel with the execution of other phases by the information architecture and user experience teams.

Content Checklist

Copy

☐ Who will provide copy?

☐ Is there a budget for a copywriter?

☐ What are the copy mandatories?

☐ What's the correct tone for the audience and brand?

Imagery & Artwork

☐ Is there existing imagery?
If yes, what format and resolution is it?

☐ Is there a budget for a photo shoot?

☐ Is there a stock photo budget?

☐ Are any custom illustrations needed?

Code

☐ What code can be reused, if any?

☐ Does this require custom programming or an off-the-shelf solution?

☐ Will there be a content management system (CMS)?

☐ Who will manage the content?

File Organization & Naming

A designer's ability to organize his or her working and production files is always important, but with web design it's critical. This is because the files that a designer uses to create a site are the same files that a user will download and view on his or her computer. Factors such as file name, file type, file size, and directory organization are all significantly more important than with print design. HTML files reference other files with relative paths, which means they find other files based on their own location. Therefore, files need to be organized in clearly labeled directories, as seen in the diagram to the right.

Properly naming files can help improve workflow and, more importantly, ensure the files will be handled properly by the web server. Rule number one is never use spaces in file names. While Mac and Windows systems can handle spaces with no issue, servers running UNIX can have difficulty with spaces.

Clear file names help the programmer understand the content of the file and they help organize the directories for a website. The example file names seen here are all buttons, thus they start with "btn_" and because of this they group together alphabetically. Note that they're all lowercase as well. This is for consistency and because some languages like XML and XHTML are case sensitive, so to be safe designers should stick with an all-lowercase convention.

Main Directory

HTML/CSS files

Image directory

Javascript directory

Media directory

btn_red.png
btn_blue.png
btn_green.png
btn_orange.png

"Logically" can mean a number of things:
Logically from a **business point of view**;
or logically from a **user's point of view**.

Taxonomy and Grouping

Once the objectives have been set and all of the things the client would like to say
and do with the site have been established, you can begin organizing and mapping
out the content. Start by listing all the content. Then begin grouping the content
logically. "Logically" can mean a number of things—logically from a business point
of view (or how the client sees things being grouped), or logically from a user's
point of view (or how the information will be consumed).

Some information architects conduct this exercise with software like OmniGraffle,
or the old-fashioned way, with index cards. How you decide to do this is up to you,
but the ultimate solution should be something that makes sense for the user and
the client.

One rule of thumb is to limit the number of choices a user sees at any given point
to seven items. Physiologically, humans cannot perceive more than seven items
at one time without creating subgroups. So to not overwhelm the user, a typical
primary navigation will not have more than seven items, often fewer.

Hierarchy of information can begin to unfold with the selections for the primary
and secondary navigations. The primary navigation should be just that, the primary
activities that a user will want to conduct on the site. The secondary navigation,
which often has less visual importance, is for supporting content.

There is a constant tension between the desires of the client and the needs of
a user. A good example of this is the "About Us" link found on many company
websites. As a business owner (and I have seen this many times), there is a
temptation to put this first in the navigation — "I want them to know our story!" the
client will often say. However, from the user's perspective, "About Us" is often the
last thing he or she is looking for on a website. Using the tools previously discussed
in this chapter — briefs, SWAT charts, etc.—are a great way to provide objective
reasoning to a client to present a clear hierarchy of information for the user.

Information Architecture

In order to understand what information architecture (also known as IA) is, it's important to unlearn what most designers think IA is. It is a mistake to think of IA as simply a means of sketching a design—boxes and shapes that represent the "underpainting" of a layout. While this may be useful to some designers and may also be how IA got its start, it's only a sliver of the IA field, which extends well beyond simple design planning.

In the infancy of the internet, websites were predominantly "information spaces"—news sites, medical sites, marketing brochure sites, etc. Therefore, there was a need to "architect" these spaces, which meant designing effective ways to a) organize the content and b) navigate through it so users could easily find what they were looking for. If you look at the deliverables an information architect created, that becomes clear.

 Site maps: An illustration or map of the pages of a site and their relationship to one another

 Taxonomies: The classification of content into a hierarchical structure

 Labeling systems: The process of naming buttons and links to make it clear what content they will reveal

 Wireframes: A means of organizing the content of an individual page as well as illustrating any technical requirements needed

These tools illustrate the navigation structure and provide context as well as the details of the various information components shared across the different pages/screens of the site.

When websites became more transactional, IAs started to become more like interaction designers, thinking in terms of discrete user tasks, mapping out user flows, designing—from a functional point of view—the individual components that would allow users to complete tasks, and all the nitty-gritty that went into each component. At this point, IAs shifted their attention away from the client and/or the designer and focused it squarely on the user. User scenarios became a standard IA deliverable, showing the paths to desired user outcomes and (usually) business outcomes, too.

Sitemap

The information architecture phase of a website project starts with the development of a comprehensive sitemap. A sitemap is a schematic for a site showing the pages and the linked relationships among them. Traditionally, pages are represented by outlined boxes, and links are represented by lines connecting the boxes. This document gives a design team an overview of the site and allows designers to understand the breadth of the navigational needs and the full scope of the project: What pages are most important? What pages need to be reached from every page? Is there a target page that the client wants to lead people to? All of these questions can be answered by examining a sitemap.

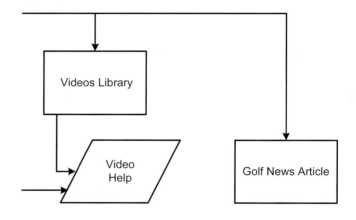

A sitemap, like the one shown here for GolfersMD.com, shows the pages of a site and their relationship to one another. Pages are laid out and grouped by the information architect, showing various pathways and connections that a design team uses when laying out the navigation and sub-navigation. In this case, the items in the main navigation are shaded in blue, pages that require the user to log in are shaded in gray, and pop-up windows are slanted boxes.

The sitemap on the next page illustrates that even a very large and seemingly unwieldy site becomes more manageable when neatly organized by an information architect.

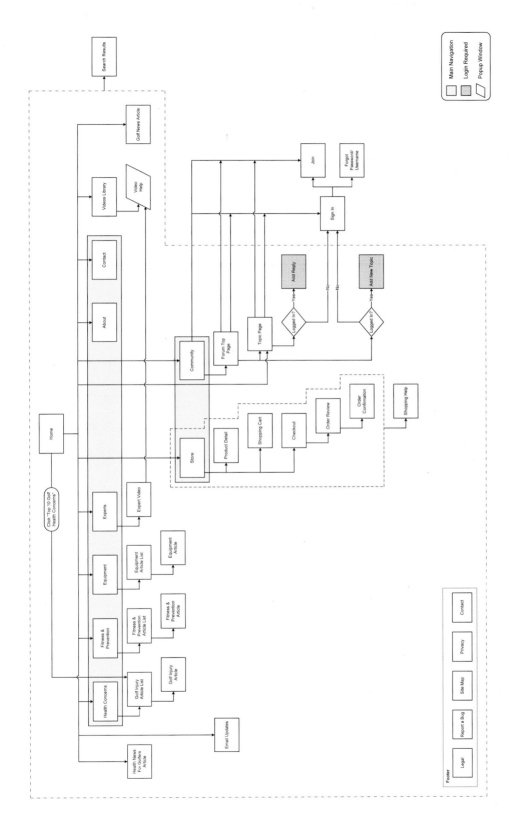

Search Results

Golf News Article

Videos Library

Video Help

Contact

About

Join

Forgot Password/Username

Sign In

Community

Forum Top Page

Topic Page

Add Reply

Add New Topic

Logged In? — Yes
No

Logged In? — Yes
No

Home

Click "Top 10 Golf Health Concerns"

Store

Product Detail

Shopping Cart

Checkout

Order Review

Order Confirmation

Shopping Help

Experts

Expert Video

Equipment

Equipment Article List

Equipment Article

Fitness & Prevention

Fitness & Prevention Article List

Fitness & Prevention Article

Health Concerns

Golf Injury Article List

Golf Injury Article

Email Updates

Health News For Golfers Article

Footer

Legal

Report a Bug

Site Map

Privacy

Contact

Main Navigation

Login Required

Popup Window

32

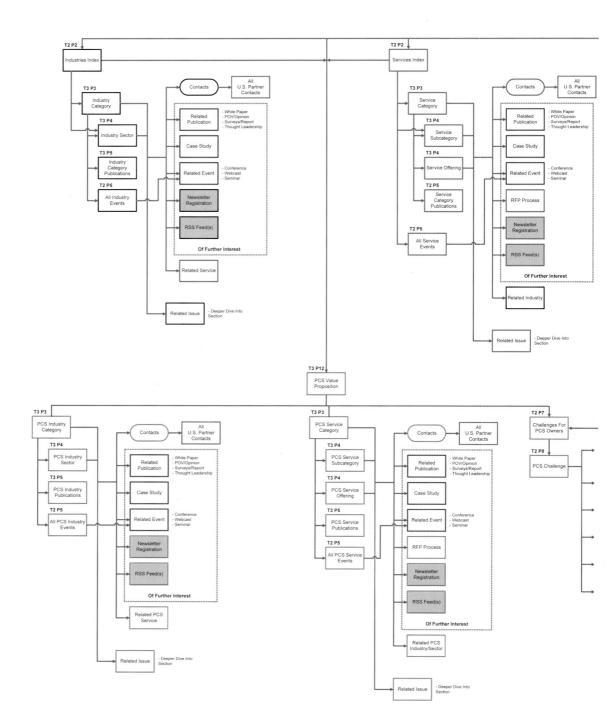

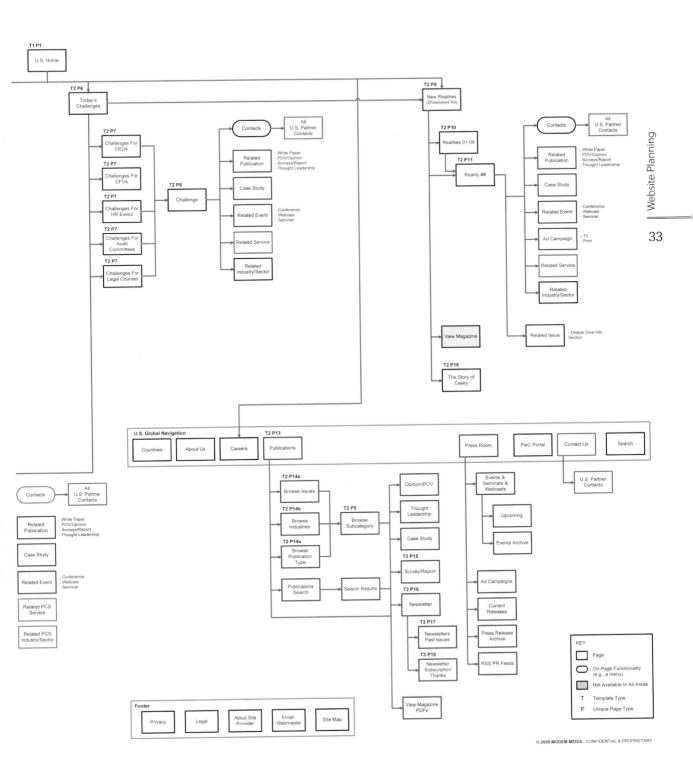

T1 P1
U.S. Home

T2 P6
Today's Challenges

T2 P9
New Realities (Showcased Ad)

Website Planning

33

T2 P7
Challenges For CEOs

T2 P7
Challenges For CFOs

T2 P7
Challenges For HR Execs

T2 P7
Challenges For Audit Committees

T2 P7
Challenges For Legal Counsel

T2 P8
Challenge

Contacts

All U.S. Partner Contacts

Related Publication
- White Paper
- POV/Opinion
- Surveys/Report
- Thought Leadership

Case Study

Related Event
- Conference
- Webcast
- Seminar

Related Service

Related Industry/Sector

T2 P10
Realities 01-09

T2 P11
Reality ##

Contacts

All U.S. Partner Contacts

Related Publication
- White Paper
- POV/Opinion
- Surveys/Report
- Thought Leadership

Case Study

Related Event
- Conference
- Webcast
- Seminar

Ad Campaign
- TV
- Print

Related Service

Related Industry/Sector

View Magazine

Related Issue
- Deeper Dive Into Section

T2 P19
The Story of Casey

U.S. Global Navigation
Countries | About Us | Careers | Publications **T2 P13**

Press Room | PwC Portal | Contact Us | Search

U.S. Partner Contacts

Contacts

All U.S. Partner Contacts

Related Publication
- White Paper
- POV/Opinion
- Surveys/Report
- Thought Leadership

Case Study

Related Event
- Conference
- Webcast
- Seminar

Related PCS Service

Related PCS Industry/Sector

T2 P14a
Browse Issues

T2 P14b
Browse Industries

T2 P14a
Browse Publication Type

Publications Search

T2 P5
Browse Subcategory

Search Results

Opinion/POV

Thought Leadership

Case Study

T3 P15
Survey/Report

T3 P16
Newsletter

T3 P17
Newsletters Past Issues

T3 P18
Newsletter Subscription Thanks

View Magazine PDFs

Events & Seminars & Webcasts

Upcoming

Events Archive

Ad Campaigns

Current Releases

Press Release Archive

RSS PR Feeds

KEY
- ▢ Page
- ⬭ On-Page Functionality (e.g., a menu)
- ▨ Not Available In All Areas
- T Template Type
- P Unique Page Type

Footer
Privacy | Legal | About Site Provider | Email Webmaster | Site Map

© 2005 MODEM MEDIA   CONFIDENTIAL & PROPRIETARY

Wireframes are **blueprints** that map out individual pages of a site. They show the elements of a page and their relative weight or importance.

Wireframe

Wireframes are blueprints that map out individual pages. The wireframe shows the elements of a page and their relative weight or importance. They are not intended to illustrate the layout of the page; instead, they visually catalog the elements on a page and give a designer an idea of what the most important elements are, what the second most important elements are, and so on. They can also detail specific functionality for a page; illustrate different states for elements on the page or the entire page, like drop-down menus or expanding areas; or demonstrate how modular areas might work together.

Wireframes can be made for any page of the site that requires this type of detail, like the home page, subpage templates, registration forms, search results, and so on. This step helps a designer focus on style rather than a dual task of form and function during the layout phase.

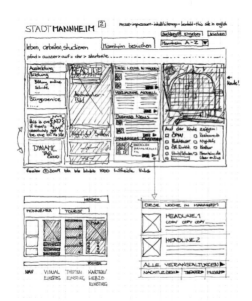

The wireframes seen here and on the next spread are what user experience experts and information architects use to organize a page for a design team. Wireframes are the bones that a designer uses to flesh out by adding brand elements and aesthetic treatments.

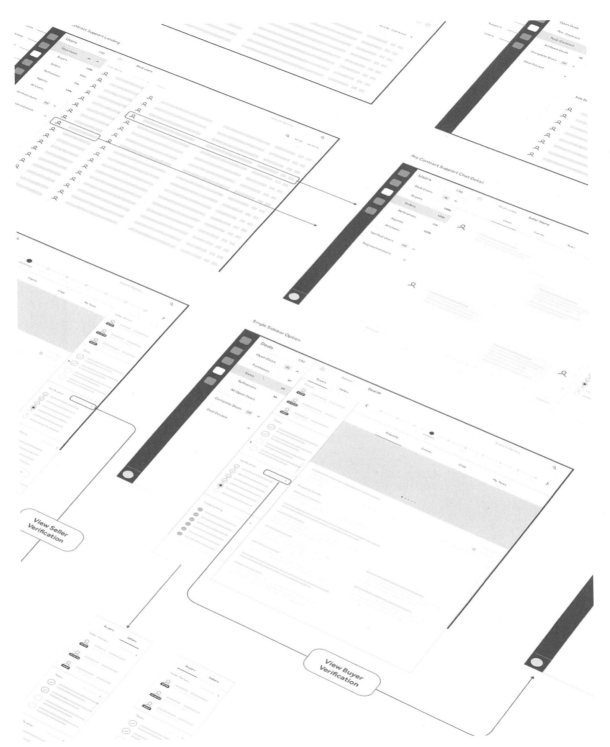

View Seller
Verification

View Buyer
Verification

Welcome

Lorem ipsum dolor sit amet,
consectetur adipiscing elit, sed do
eiusmod tempor incididunt ut labore
et dolore magna aliqua

Email

Password

SUBMIT

Explore Funds

LOAD MORE

Bonaccord
Capital Partners

Minority Interests in Mid-Sized
Managers– $1,000,000,000

REQUEST INFORMATION

FOLLOW

Bonaccord
Capital Partners

Data Room

Contact Eaton
Marketplace

Lorem ipsum dolor sit amet,
consectetur adipiscing elit

Name

Email

SUBMIT

Privacy Policy

Usability Diagrams

Usability diagrams (also known as user-flow diagrams or use cases) combine a sitemap and a wireframe to plan out a specific action a user might take on a site, and the process of how it occurs. Each step of a process is illustrated especially for tasks that have multiple outcomes, like error and success messages. For example, to show how someone might register as a user on a site, a usability diagram would show a home page, a registration page that's linked from the home page, an error page showing that the user didn't complete all the required fields, a "thank you" page showing the registration was complete, and a confirmation email wireframe. User-flow diagrams show every step of the process and can help uncover potential issues. The process of creating a use case can be as valuable as the resulting diagram. The exercise of acting as a user and imagining interaction with the site is a critical preparation step in designing for the web.

The usability diagram seen here goes a step beyond a sitemap and illustrates the path a user might take through a site. The diagram can include not only on-site pages, but emails and even off-site actions like going to a retail store or calling an 800 number. These help the web project team lead the user to the intended goal of the client in the most effective way.

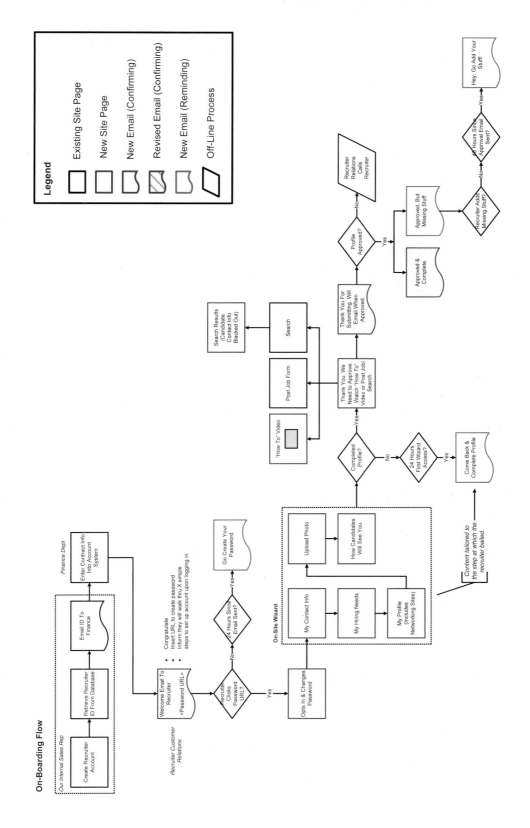

Website Planning

Prototypes are working models of site functionality that help a developer work out the final details and provide **proof of concept.**

Prototypes

Once the wireframing is complete and critical tasks are mapped out, it's sometimes necessary to create functional prototypes for new or complicated functionality. Prototypes are working models of site features or functionality that help a developer and a designer work out the final details and provide proof of concept. These working models, which are usually void of any design treatment, provide valuable opportunities for evaluation and refinement that can't be done with diagrams alone. Once a prototype is functional, it's ready to be "skinned" by the designer. Skinning is a term used by designers that means to add a design treatment on top of a working model.

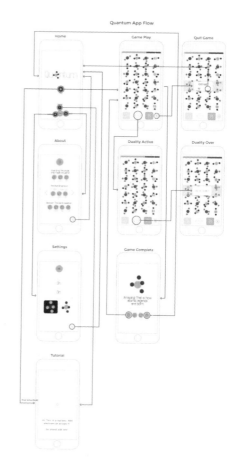

Prototypes, like the ones seen here, are created by the development team to flesh out specific technological challenges and to create a proof of concept that an idea can actually be executed.

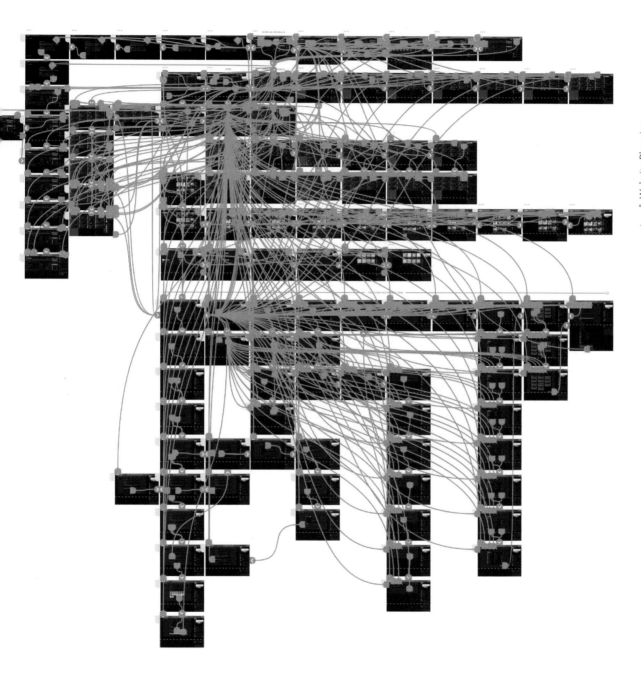

Concept Design: Mood Boards

The beginning creative stages of a web design project are not unlike other creative projects. They involve understanding the goals of the client, understanding the audience, and creating a vision for how those two ideas can meet—and of course, how the designer can express creativity and originality in the process.

One way that designers begin the creative process is to gather and collect visual samples that relate to the visual feel or brand image for a specific project. These visual samples, or swipe, can come from anywhere—sites like Pinterest.com are a great place to start, but designers also pull from physical magazines and catalogs, Google image searches, and by taking their own photographs. The swipe collected for a mood board can include:

Imagery: Finding the right imagery style for a project can help the designer understand the creative direction.

Iconography: Iconography styles vary widely. A mood board should present a single, consistent style.

Color: Having a color theme for a project is important for providing visual unity and setting a tone or mood.

Texture: Often overlooked, texture can bring a concept to life and add a richness that sets a design apart.

Typography: Type plays a critical role in any design and should therefore be carefully considered in relation to the other design elements.

Seen above is a Pinterest.com board of type, imagery, and illustration, which will serve as inspiration for a design project. Pinterest is a great way to quickly collect and organize elements of a mood board.

The mood boards seen here and on the next page were created by the Wonderfactory to help their client get a feel for the visual mood of a site prior to seeing the finished design, also shown here.

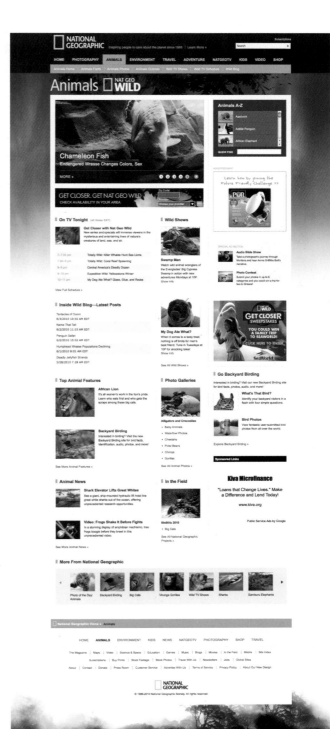

To create an effective mood board, designers start by grabbing as many appealing samples of these elements as they can find, collecting them in cohesive groups. Telling a story is a big part of this exercise, but simply grouping elements based on the story they tell isn't enough; there must be visual unity. Once you have a visual story with several images, colors, texture, and typography, codify them onto a mood board.

Mood boards help you and the client understand the overall look of a site without getting bogged down with the details of navigation or other details of a Website. The loose feel lets clients use their imagination—with a little help from you, of course. Most of all, a mood board is meant to garner an emotional response from the client—"I love it!" In addition to helping your client understand your vision for a site, a mood board can also help you brief your team and focus their energy in a particular creative direction. In either case—briefing or presenting—the purpose of a mood board is to be shown early. The items of a mood board, while illustrative, should be very easy to change if the client hates it.

Style Tiles

After presenting and gaining approval on a mood board, the client might require a greater level of detail before moving forward with the design. For this web designers create what are known as style tiles. Style tiles provide a higher level of detail to the visual story, but still fall short of an actual layout of a web page.

As the name implies, style tiles go beyond simply showing an array of elements; they depict specific styles that might be applied to the design elements. For example, a style tile might show a stroke or shadowing effect around photography, how dimension might be used, or how specific type treatments like headlines or body copy might appear.

Information architecture, mood boards, and style tiles are all means of building the experience that is right for a particular target user, and they help designers avoid falling into the cookie-cutter generic design style that is seen all over the web. The colors, photos, textures, type, etc., that have been established with the mood boards and style tiles are the foundation for the user's experience.

Metaphors

46

What mood boards and style tiles help to do is establish a metaphor for a site. A metaphor is defined as "a thing regarded as representative of something else." When designing and interacting with web pages, it's easy to forget that there aren't actually "buttons" or "tabs" that users "press." Those are just metaphors from the real world of dashboards, calculators, file folders, etc., that have developed into a visual explanation of a clickable item.

Metaphors make the unfamiliar familiar. They take abstract ideas, like linking text from one page to another, and make them tangible. They help users relate to the content and the design of a site. The right metaphor can help reduce the need for instructional copy by creating a setting or environment that is familiar to the audience.

Ultimately, finding the right experience for your user is what designers strive to do. Creating mood boards, style tiles, and developing a visual metaphor are ways in which designers create sites that unique to their clients' brands and right for the desired experience of the target user audience.

The next chapter explores turning these metaphors into meaningful experiences for your target user.

Seen here is the next step in the creative development process. Once the mood board is established, the interface elements can be created in the approved style.

(Opposite) This famous painting by Magritte (c.1929) can remind us that when we're surfing on the web there are no buttons to be pushed—only pixels on a screen.

Ceci n'est pas une pipe.

CHAPTER 2

Elements
of Usability

To effectively plan out a Website project, a designer must have a good understanding of usability. Usability is a term that refers to the ease with which users can learn, engage with, and get satisfaction from an interface for a website or piece of software. While the IA documentation, like usability diagrams, is helpful for a designer in planning out a website, usability effectiveness also comes from a variety of other factors—design, server speed, technology usage, animation, and even sound effects. This chapter explores the following interface elements, which, when combined, cover the usability touchpoints for a user: navigation, breadcrumbs, site search, submission forms, links and buttons, and error messages. While usability comes from more than just these interface elements, these are the features of a site that a designer can most greatly influence.

Enough About You

Clients   Users

This humorous cartoon illustrates the differences in perception between the client and the user. And it highlights the need for a designer to put themselves in the position of a user to understand how they perceive the product they're working on.

Usability is about the user (period). Usability is directly related to the experience a user has with a site—the better the usability, the better the experience is likely to be. Individual users vary widely, even within a single target market. In web design, standard demographic data such as age, education, gender, language, interests, and culture apply exactly as they do in other forms of communication—but there's an added level of demographic information that includes technology, like operating system, processor speed, screen resolution, memory, and network connection speed. All of this demographic information can play an influential role when it comes to usability design.

Usability is such a critical aspect of web design that many web design agencies employ user experience (UX) experts. Part sociologist, part technician, this person is responsible for determining the most appropriate usability based on the abilities and expectations of the target user group, as well as the technology that's available. Some of the factors that usability experts consider include:

PAGE LOAD SPEED: For desktop, this is less of an issue, since most computers are connected to the internet with high-speed connections. But for mobile, this is essential. Not only will pages load more quickly, but the site will use less of the user's mobile data allowance.

LEGIBILITY: In all cases, the legibility of the typography—including adequate contrast between the type color and the background, sizing, line spacing, and font choice—is essential for increasing the usability of a site. There will be more about this topic in chapter 6.

ACCESSIBILITY: Over the past several years, accessibility of websites has become a central consideration for web designers and developers. More than simply adding ALT tags so images can be described for the visual impaired, accessibility covers a wide variety of visual and coding factors. To learn more about the Americans with Disabilities Act requirements, visit ADA.gov.

SCANNABLE CONTENT: Users come to a site for content, plain and simple. So the content is a key part of the usability of a site. Content should be broken up into manageable bits with descriptive headings, making pages easily scannable by the user.

CLEAR URLS AND PAGE TITLES: The page title appears in the header of a web browser and it tells the user what the content of the page is. It also tells search engines what the content of the page is. Accurate and clear page titles help users find the right content.

CONSISTENT DESIGN TREATMENTS: The design of a site needs to hold together and be consistent for the user to be able to recognize various elements of a page. This is also true about the mobile experience of a site. It should share consistent design treatments with the desktop and vice versa.

CROSS-CHANNEL USABILITY: More than ever, users start browsing for content on mobile and tablet devices. Therefore, UX and web designers must consider the mobile experience as a critical part of the overall experience of a website. The principles of usability that follow are universal; however, the specific design treatments may vary on mobile to increase usability. Wherever possible, examples of desktop and mobile samples are shown.

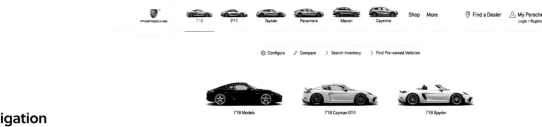

Navigation

Navigation is a broad term that encompasses any aspect of a site that links a user to another area of the site and is the cornerstone of a site's usability. Unlike other forms of information design that have a natural sequence—pages of a book or brochure, for example—web pages present users with a menu of options and allow them to choose their own order. The main navigation of a site is the primary set of links that a user clicks to get to the important content of a site. The most common convention for main navigation is a persistent bar across the upper part of a page that features a list of five to seven options, with other options relegated to sub-navigation. (Groups of five to seven are generally what people are capable of perceiving before attempting to break them down into subgroups.)

There are two ways of dealing with large site architectures: Categorize content into main sections, then use a cascading system of menus either with drop-down lists or sub-menus; or break up the list of choices into the most important items (primary navigation) and the lesser important items (secondary navigation). In either case, six groups of five are much easier to comprehend than one group of thirty. Either method makes comprehending the site architecture easier for the user and reduces the number of clicks it takes for a user to get from one place on the site to another.

This photo-driven drop-down menu on Porsche.com expands to make finding content very easy for the user.

Thanks in part to the rise of mobile responsive design, these three lines, known as a hamburger, have ubiquitously become the symbol for navigation or menus on websites.

*These images show
the drop-down
menu navigation on
newyorkmag.com.
Information is grouped
into six main categories
for greater usability.*

INTELLIGENCER | THE CUT | VULTURE | THE STRATEGIST | CURBED | GRUB STREET | MAGAZINE ⌄ | THE CITY   Subscribe | Sign In

Flash sale: Save 60% today

New York

PROFILE
Esther Perel Goes Off Script
She became today's most famous couples therapist by ignoring all the rules of the trade.
BY MAYA BINYAM

EDUCATION
Richard Carranza's Last Stand As NYC Schools Chancellor
De Blasio hired an "equity warrior." Parental politics and the pandemic left him defeated.
BY CLARE MALONE

BUSINESS
America's Failure Is Amazon's Success
Alec MacGillis's *Fulfillment* is a portrait of trickle-down inequality.
BY SARAH JONES

LATEST NEWS

What to Know About the Johnson & Johnson Vaccine Pause
The FDA and CDC recommended a temporary pause in use of the vaccine while they probe several cases of recipients experiencing severe blood clots.

What We Know About the Knoxville High School Shooting
One student was killed in an officer-involved incident at Austin-East Magnet High School on Monday.

Authorities Say Cop Accidentally Shot and Killed Daunte Wright
A state agency identified the officer who killed Wright as Kim Potter, the president of the Brooklyn Center police union.

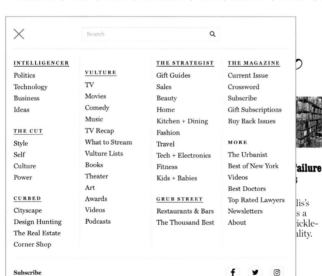

INTELLIGENCER | THE CUT | VULTURE | THE STRATEGIST | CURBED | GRUB STREET | MAGAZINE ⌄ | THE CITY   Subscribe | Sign In

Flash sale: Save 60% today

Search 🔍

INTELLIGENCER
Politics
Technology
Business
Ideas

THE CUT
Style
Self
Culture
Power

CURBED
Cityscape
Design Hunting
The Real Estate
Corner Shop

VULTURE
TV
Movies
Comedy
Music
TV Recap
What to Stream
Vulture Lists
Books
Theater
Art
Awards
Videos
Podcasts

THE STRATEGIST
Gift Guides
Sales
Beauty
Home
Kitchen + Dining
Fashion
Travel
Tech + Electronics
Fitness
Kids + Babies

GRUB STREET
Restaurants & Bars
The Thousand Best

THE MAGAZINE
Current Issue
Crossword
Subscribe
Gift Subscriptions
Buy Back Issues

MORE
The Urbanist
Best of New York
Videos
Best Doctors
Top Rated Lawyers
Newsletters
About

Subscribe   f 🐦 📷

LATEST NEWS

What to Know About the Johnson & Johnson Vaccine Pause
The FDA and CDC recommended a temporary pause in use of the vaccine while they probe several cases of recipients experiencing severe blood clots.

What We Know About the Knoxville High School Shooting
One student was killed in an officer-involved incident at Austin-East Magnet High School on Monday.

Authorities Say Cop Accidentally Shot and Killed Daunte Wright
A state agency identified the officer who killed Wright as Kim Potter, the president of the Brooklyn Center police union.

Navigational elements need to visually stand apart from the rest of the elements on the page and indicate that the user can click on them. There are usually four states to an item in a navigation bar: the dormant or static state; the active state, which indicates the current page; a rollover state, which is sometimes the same as the active state when a user mouses over the button; and the visited state, which indicates to the user what's already been visited. This system should be easy for the user to learn and should remain consistent throughout the entire site.

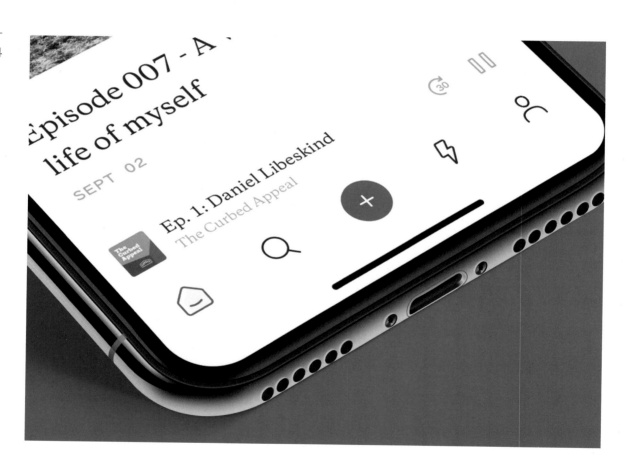

Another type of navigation that is popular on mobile sites and apps is called Bottom Bar Navigation. This area of the screen is more accessible with a user's thumbs on a hand-held device.

The language of a button should clearly and accurately predict the content of the destination page. The labels should be written from the user's perspective, with terms users might use to find what they're looking for. (Users are quick to abandon a site if they have been confused or deceived by a misleading button.) In addition, since search engines often value the text within links, it's important to use keyword-rich terms in the navigation. This is also why the most effective navigation bars use web fonts for the buttons—not images of text, which are unreadable by search engines.

The topic of navigation is explored further from a design perspective in chapter 4, "Anatomy of a Web Page."

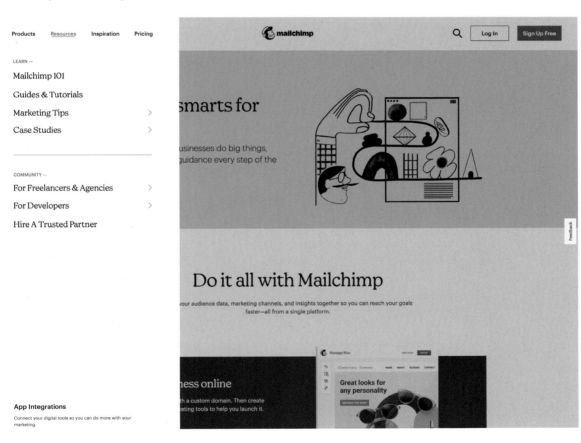

This lovely example from MailChimp.com uses a combination of three different styles in a single navigations system: horizontal nav, vertical or left-side nav, and the mega menu.

Breadcrumbs

A useful subset of navigation is something called breadcrumb links or breadcrumbs. Generally located at the top of a page below the header, breadcrumb links reveal to the user the path taken through the site architecture to get to the current page. Breadcrumbs make it easy to retrace your steps and get back to a previous page should you find yourself on a page you no longer have use for. The name breadcrumbs comes from the story of Hansel and Gretel, when Hansel scattered crumbs of bread on the ground to help him and his sister find their way home. Unfortunately for the pair, birds came along and ate their breadcrumbs, but the metaphor lives on as a trail of tasty links guiding users on websites.

A form of breadcrumbing is also used for submission forms. An indicator bar is sometimes used across the top of a form to reveal the number of steps in the process—both what they've completed as well as the steps yet to come. This helps estimate how long a submission form is and whether it's worth the user's time to complete.

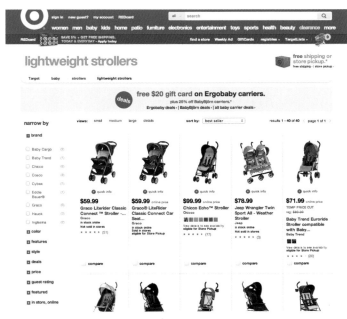

Breadcrumb links like those seen in these samples act as a sub-navigation that lights a user's way back to the home page.

Breadcrumbs can be simple links or more elaborate drop-down menus like the ones seen above. In both cases they help the user ground him or herself on the site.

Links

Within the content of a site, it's often necessary to link users to other areas of the site for additional content. This granular level of navigation is helpful to users who want to know more about a specific idea, and helpful for SEO because linked words have high indexing value. Since linked text usually consists of keywords from the article, highlighting the links helps the "scannability" of a page—a user can scan and read the linked words and get a general sense for the content of the page. Links in long bodies of text, however, can also be a distraction to a user who's trying to focus on a single story. For this reason, links should stand out so they can be recognized, but not so much so that they're distracting.

According to leading usability authority Jakob Nielsen (useit.com), the best method for indicating a text link is underlining and changing the color of linked text; however, any alteration is available when indicating a link in CSS. Aside from the indication of a link, there should also be two other visual states of a link: mouse over, and visited. The mouse-over state gives the user visual feedback that the text is indeed a link and not just underlined for emphasis. The visited state helps the user recognize where he or she has been. There's also a less common active state, which appears the moment a user clicks.

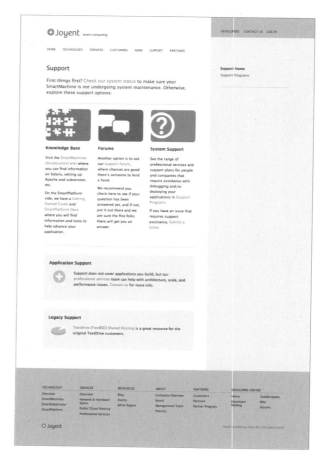

These examples from AndyRutledge.com (top) and JustWatchTheSky.com (below) show alternate ways to highlight links. Any CSS style variation is possible when indicating links, from underlining and color changes to size, weight, and background color shifts.

Buttons & Sequencing

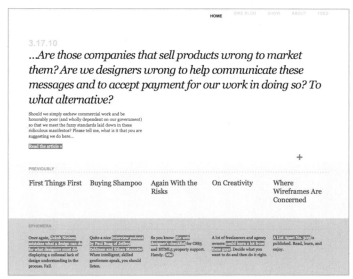

Buttons play a major role in web design and help guide a user through the content of a site. Like links, color theory plays a major role in the successful usability of a button. For example, a solid green button generally indicates that the user will advance, while an outlined or white button can often mean cancel or go back. Similarly, a button on the right side of the screen will indicate to a user advancement since in Western culture we read from left to right. And a button on the left side of the screen can indicate regression.

Site Search

Perhaps the quickest way to allow users to find information on a website is through a site search feature. Search forms search a database of site content and display the results for a user, linking them directly to the item they came for—ideally. Because a search box is intended to increase usability, it should be as easy to find and use as possible. This means placing it above the fold in a conspicuous location that's consistent on every page and clearly labeled "Search" or something similar. Also, it's important to make the search field long enough to accommodate the types of searches people will conduct. Although longer search terms can be entered into a short field, users tend to edit themselves if they're given a small space. It is also possible to pre-populate the search form with the type of search available through the form.

Internal site searches will sometimes have an advanced search feature. This is an extension of the search functionality with added fields that allow a user to narrow down a search to increase the likelihood of finding what is needed. The most effective search boxes have the ability to remember popular searches and match them to the characters entered by the user so the user can see, then click on, a list of potential search terms and be redirected to those results.

The search features on Typography.com, which includes "find fonts" and "browse collections" drop-down features, make finding content on the site easy and intuitive.

Usability Testing

While creative focus groups can be the death of fresh ideas, usability testing, which consists of inviting potential users to complete a series of tasks using the interface concept, can greatly help refine the usability elements of a site. During a usability test session, the moderator observes and records the users' reactions and emotions as they attempt to complete a given task. Confusion or frustration expressed by the user help pinpoint trouble spots, whereas delight or satisfaction means that the usability is appropriate for the task and the user.

To the right is a sample transcript from a usability test. In this example the subject tester is asked to find books about graphic design. The moderator prompts the user with tasks and nothing more. The user's actions and quotes are recorded and the icons indicate positive or negative feedback, as well as feedback that represents an idea by the user.

FIRST IMPRESSIONS
"I like the design and the colors, but I don't know where to begin. I suppose if I had something to do here I would know where to start."

PLEASE SEARCH FOR INFORMATION ABOUT GRAPHIC DESIGN BOOKS.
User starts search

"The search field is a bit short, which makes me think I can only search for single terms."

User receives 18 results.

"It would be great if these results could be sorted by price and availability."

User really likes the layout of the results page, including the thumbnail images of the books.

PLEASE SELECT A BOOK FOR PURCHASE
User clicks the thumbnail of the book to view detail and nothing happens.

"I should be able to click the image of the book to see the product descriptions."

Subject clicks "Learn More" and sees product description page.

"I like this page, but it's too hard to find the price. I want to know immediately how much this book costs."

User adds book to shopping cart.

"I like how I don't leave the page when the book is added to the cart."

User clicks the "Check Out" link and proceeds to check out page.

The search field is only half of a site search solution; the **search results page** is the other.

The search field is only half of a site search solution; the search results page is the other. There are a couple of important features of a results page that can help with usability. The searched term should remain in the search box at the top of the page and the number of results found should also clearly be displayed. Effective search results pages give users the ability to sort the results—by date, by relevance, or by author, for example. The search results themselves should display enough key information so the user can make an informed decision as to whether the results are the desired ones. Finally, on the article page, there should be a mechanism that allows users to rate the relevance and quality of the article based on the user's search criteria. This will teach the search engines what content is most relevant for different search terms.

The search results on Gap.com feature photos of the clothing related to the search by the user. The search results also have a filtering feature that makes narrowing down the selections easy.

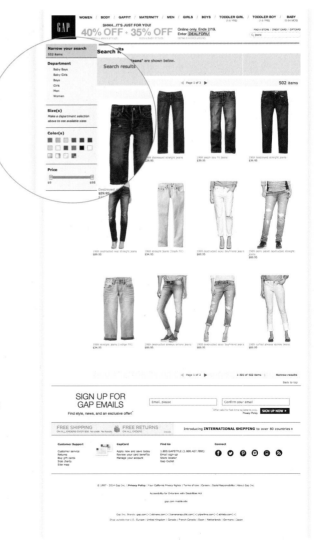

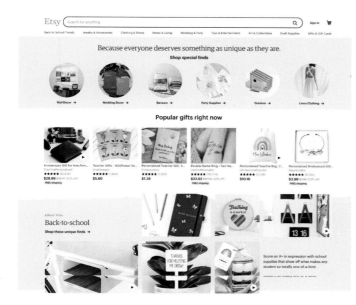

Etsy.com has a very clean top navigation with an equally intuitive search feature. The search area stretches across the entire page and features search suggestions based on the letters a user types in. The search results (bottom) are sortable and can be converted from image view to list view.

As a general rule, users don't like filling out forms, so it's the job of the designer and UX specialist to make the process as **pain-free** as possible.

Submission Forms

Submission forms, where a user inputs information and submits it to the site, generally represent a goal for a site—inviting the user to register, sign up for a newsletter, buy a product—so the usability of a submission form is of premium importance. Unfortunately, as a general rule, users don't like filling out forms, so it's the job of the designer and UX specialist to make the process as pain-free as possible. It's important to be clear about the length of the form up front, with long forms broken up into manageable segments with a breadcrumb trail indicating what's left to come.

A form is a series of fields that a user fills out with information. The fields should be clearly labeled with the information that needs to go in them. Designing the labels to the left of the field, as opposed to above them, will give the appearance of a shorter form. Required and optional fields should be indicated clearly so the user knows what fields can be skipped. Fields should be grouped in a logical way so the user can follow the flow easily, and redundant information, such as shipping versus billing information, should be pre-populated if the user desires. When validation (an available username, for example) is required, it should be given in process, not after the form has been submitted. The number of times a user has to correct errors and resubmit a form greatly increases the likelihood that the user will drop off.

CONTACT

Adding style to a submission form can make it more inviting for the user. The forms seen here, from the simple email form above to the more complex content management forms on the opposite page, benefit from a clear grid, generous white space, and typographic hierarchy.

It's often useful for a designer to limit the number of actions a user can take when on a form page. This can mean removing all global navigation and limiting the clickable options to "Submit" and possibly some "Help" links. After submitting a form, a user should be given a clear indication that the submission was successful.

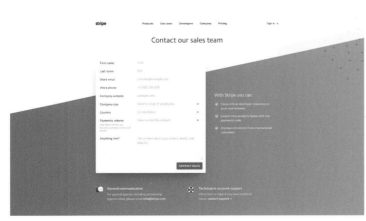

These form examples utilize an underlying grid structure to organize the space in the layout, which helps minimize the appearance of large amounts of information to fill in.

These form examples from MailChimp.com (top) and OmmWriter.com (bottom) style the form elements in a way that causes them to blend in with the design. Although rarely taken advantage of, CSS can style form fields just like any other element within a design.

This form uses numbered points to indicate the number of steps that are required to complete the form. This makes each process manageable for the user and gives them a sense of context.

Each **form element** has a specific purpose that a designer should understand when designing an online form.

Form Title

Input 1

Password

Input 2

⦿ Option 1
Single-line description copy for option 1

◯ Option 2
Single-line description copy for option 2

◯ Option 3
Single-line description copy for option 3

Selection 1

| Select | ▼ |

Selection 2

☒ Choice 1

☐ Choice 2

Action

This sample form shows the various form elements that are used to collect information. Each element has a specific purpose that a designer should understand when designing an online form.

There are three types of text fields: text box collects a single line of information; text box with password protection collects a single line of information but the user only sees bullets or asterisks; and text area, which can collect multiple lines of text. Text fields can be set to be pre-populated with a phrase to help the user understand the type of information that can be input.

For selecting items there are three main choices: radio buttons (seen as circles in this diagram) are mutually exclusive—meaning only one can be selected from a group—and they allow for written explanations of the options; drop-down menus are also mutually exclusive and they provide a simple list of items; and check boxes (seen as boxes in this diagram), which are used for allowing the user to select multiple options.

The submit button triggers the action of a form and can either be a browser-generated user interface (UI) element, an image, or text.

Error Messages

Despite the best efforts of designers and UX experts, users will sometimes come across an error on a site. The most common errors on submission forms occur when the proper information is not filled in correctly. Indicating an error clearly can be essential in converting users who are willing to spend time filling out a form. To clearly indicate an error, a designer should visually separate the error message from the page so the user easily notices it. The content of the message should be clear yet polite, and the offending form element should be highlighted clearly so the user can find it quickly and make the correction.

This example from OnSugar.com is not an error, but a hint that appears as the user selects the various form fields. This proactive approach can help reduce the need for error messaging altogether.

The sign-up form on Dunked.com starts off with an X on the username field and turns to a check when a usable name is typed in the field.

In the lower example on this page—barleysgville.com—the error message is displayed as a single line below the form, with a list of the missed fields.

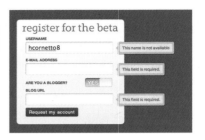

These examples range from subtle markings to obvious red fields and caution icons. The right level of strength for the error message depends as much on the layout environment it appears in as the experience level of the user group that will be using the site.

"**Something went technically wrong.** Thanks for noticing—we're going to fix it up and have things back to normal soon."

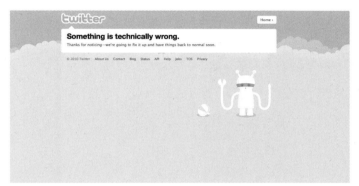

The right copy can play an important role in effective error messaging, since it's easy for the user to feel like he or she has done something wrong. In this example from Twitter.com, the copy reads, "Something went technically wrong. Thanks for noticing—we're going to fix it up and have things back to normal soon."

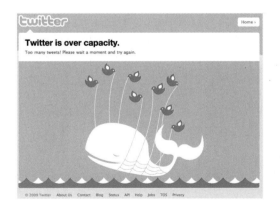

The "Fail Whale," also from Twitter.com, is a surprising yet delightful error message. The unexpected nature of a whale being flown by birds makes finding an error almost forgivable.

The surprising number of hits a 404 page receives makes it a **prime design opportunity** to direct the user and reinforce the client's brand.

Another form of error message is the "404 Page Not Found." Often overlooked by designers, this page appears when a user lands on a URL that no longer exists or never existed. The surprising number of hits a 404 page gets makes it a prime design opportunity to direct the user and reinforce the client's brand. Custom 404 pages should be somewhat apologetic in tone and present a series of links so the user can find what he or she originally was looking for. The ability to search or even report the missing page is an additional feature that can be added to a 404 page.

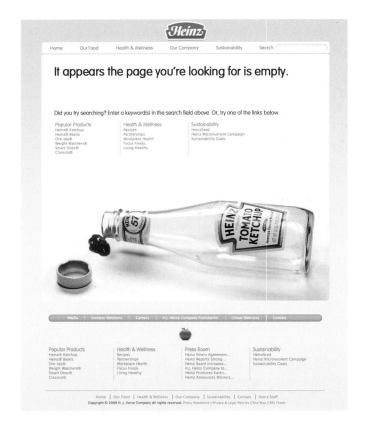

This custom 404 page from Heinz.com combines both an element of humor and the utility of being able to find information on the site.

Surprise and Delight

While the web has many utilitarian aspects to it, it's also important to remember that users—people—enjoy being entertained. "Surprise and delight" is a phrase adopted from the hospitality industry and used by web designers and UX specialists to describe the fun or unexpected features of a site. (This should not be confused with "mislead and confuse.") Surprise and delight refers to added value for a user—something that goes beyond expectations. Surprise and delight can be humorous, irreverent, or even seductive. Exactly what kind of surprise is appropriate, like anything else, depends on the target audience.

The 404 pages seen here and on the next spread are prime examples of surprise and delight.

These custom 404 pages from huwshimi.com (top) and teez.com.au (center and bottom) illustrate a sense of the company's brands, both with a sense of beauty and humor.

We looked everywhere.

And couldn't find that page. But we did find these under the couch cushions.

Not what you're looking for? Try the links below:

Personal Finance Solution
Mint.com

Personal Finance Mobile Apps
Overview iPhone iPad Android

Personal Finance Blog
MintLife.com

 Why Mailchimp?   Marketing Platform ▾   Pricing   Resources ▾   🔍   Log In   Sign Up Free

We lost this page

We searched high and low but couldn't find what you're looking
for. Let's find a better place for you to go.

Mailchimp Home

Disney   SHOP   MOVIES   SHOWS   PARKS   DISNEYLIFE

OOPS! PAGE NOT FOUND.

You must have picked the wrong door because I
haven't been able to lay my eye on the page
you've been searching for.

BACK TO HOME

Disney   SHOP   MOVIES   SHOWS   PARKS   🐦 f 📷 ▶

*The custom 404 pages on this page from
Mint.com (top), MailChimp.com (center),
and ShopDisney.com (bottom) have a sense of
brand but also provide a means for the user to
find the content he or she was seeking—from
a site nav and links to a search field.*

CHAPTER 3

Space, Grids, and Responsive Design

In the final step of the planning phase of a web design project, the designer begins to prepare the canvas for a design. This means developing a grid system that is flexible enough to accommodate a variety of content, but rigid enough to form a recognizable system. Grids are fundamentally about space, and this chapter explores means of organizing space to enhance a user's access to, and understanding of, information.

Organization and Hierarchy

One of the most important aspects of design is the concept of hierarchy. Visual hierarchy is the sequencing of elements within a design so that a user may perceive them in a specific and logical order. This sequence clearly defines the most important elements of the design, followed by the second most important elements, and so on. Almost every type of information can be broken down into three or four levels of importance. More than that makes contrasting the difference between the levels difficult.

An effective **design system** takes precedence over the individual elements, so that the user perceives a cohesive unit.

To create hierarchy, a designer must first create a system. A system is created by logically grouping the elements of a design, either through meaning or function, and forming visual relationships between them. An effective design system takes precedence over the individual elements so that the user perceives a cohesive unit. Any element that breaks this system will have more visual value and be understood to have more importance than the other elements, creating a hierarchy.

For example, in a classroom where the desks are neatly arranged in five rows of five desks and each student is sitting in his or her seat, the students appear as a single unit. Regardless of the different genders, clothing, hair styles, or body types, all the students fit within the group because of their organization or spatial relationship to one another. If a single student decided to break the system of rows by moving his desk into the aisle, he would stand apart from the system and give himself visual importance over the other students. The students appear as a single unit because of their arrangement in space—the rows of desks—and the student whose desk is not in line with the others stands out strictly because of his lack of relationship, or his contrast, with the others.

White Space

Creating a design system almost always starts with the clear organization of space. Deliberately constructed white space, not to be confused with unconsidered or empty space, is often overlooked as an element of web design. In fact, a common mistake among inexperienced designers is to focus too heavily on the "objects" in a design (type, images, points, lines, and planes), and space is simply what's left over when they're finished. Space is essential for creating relationships that form systems that lead to a clear hierarchy of elements. It should not be underestimated.

The interplay between the objects of a design and the background is called the figure-ground relationship. White space, also called negative space, is a reference to the "ground" in "figure-ground." The goal of a designer is to achieve a balance between figure and ground, where one doesn't completely dominate the other. Instead, they work together to unify the design.

White space design elements include: margins, the area surrounding a design; gutters, the space between columns of a grid; padding, the area around an element contained by a border; line spacing, also known as leading, the space from baseline to cap height between lines of text; and paragraph spacing, the space between paragraphs or separate ideas in a piece of text. Adding line space is the most common form of paragraph indication in web design, although it is possible to use other methods like indenting, which is also another form of white space utilization.

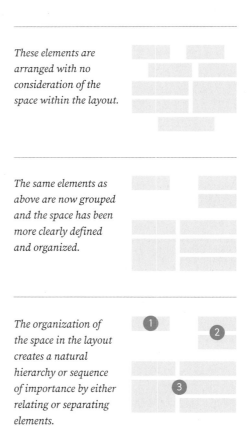

These elements are arranged with no consideration of the space within the layout.

The same elements as above are now grouped and the space has been more clearly defined and organized.

The organization of the space in the layout creates a natural hierarchy or sequence of importance by either relating or separating elements.

The Gestalt Principles of Perception:
"The whole is greater than the sum of the parts."

Theories involving the psychology of visual organization within art and design come mostly from the Gestalt Principles of Perception. These principles, developed in the early twentieth century at the Staatliches Bauhaus in Germany, refer to the mind's ability to group elements based on one of the following relationships:

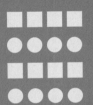

SIMILARITY
Grouping of elements that have a unique visual relationship. The two rows of squares above are grouped, despite being separated by a row of circles. The relationship of shape takes precedence over the spacial relationships.

PROXIMITY
Grouping of elements that are close to one another. Two groups are perceived above, despite the fact that there are sixteen individual boxes.

CLOSURE
Grouping of elements that complete a larger unit. A single square is perceived in the above illustration, despite several of the smaller units being removed. The small square in the upper right "closes" the spacing to create a single form.

CONTINUANCE
Grouping of elements that complete a pattern or progression. Each row of boxes above forms a group despite the gaps in the row.

Deliberately constructed white space, not to be confused with leftover, unconsidered, or empty space, is often overlooked as a useful element of web design.

InformationArchitects.jp (opposite) uses a minimalist design that relies heavily on the use of white space to organize information and create hierarchy. The gutters, line spacing, and paragraph spacing are carefully crafted to help the user identify individual groups of information.

Similarly, JonTangerine.com (this page) uses wide margins and ample padding to make the page design scannable. With the exception of a small dot of yellow and a bit of red at the bottom, this black-and-white layout uses only a single font (Georgia) yet it has a clear hierarchy of information and plenty of visual interest.

Elements of a web design aren't just design elements, they're the **interface** that the user needs to navigate and find information

The use of hierarchy and white space in web design has a bit of extra significance over their use in other forms of communication, since the elements of a design aren't just elements, they're the interface that the user needs to navigate and find information. The primary navigation bar, for example, needs to be immediately identifiable as such, so that the user can navigate the site. The design conventions discussed in the previous chapter help the user identify specific areas of a website, but they shouldn't be taken for granted. Guiding the user through a layout should be done deliberately to ensure maximum usability.

The example shown here from SnowBird.com uses white space around the imagery to mimic the branding and give the page an "ownable" feature.

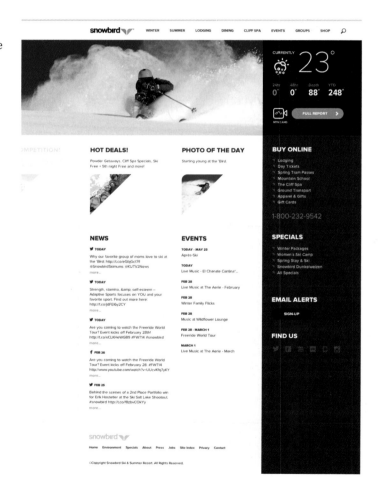

PRODUCTS   STORY   PARTNERS   IMPACT   mettā   JOURNAL   STOCKISTS   CONTACT   BUY ONLINE

Mettā Skincare is a range of natural skincare products created for you and your skin with an environmental and social consciousness.

Learn more about Metta Skincare ⌄

This sample, MettaSkinCare.com, uses generous white space to give the page, and the product, a premium feel. Nothing is cluttered; every design element is given "air."

ABOUT

Mettā Skincare is a range of natural skincare products created for you and your skin with an environmental and social consciousness. Using 100% natural ingredients and established relationships with local and international artisan producers, Mettā Skincare is more than just a product range. It is a conscious lifestyle choice that will leave you feeling good about your skin and the earth.

STORY   PARTNERS   IMPACT

LATEST JOURNAL POSTS

JANUARY 17 – 2014
Looking After Your Skin In Summer

NOVEMBER 28 – 2013
Why Natural Skincare

NOVEMBER 19 – 2013
Guide to Naturally Beautiful Skin

NOVEMBER 19 – 2013
Getting the most out of Mettā Skincare products

White space is also essential for making a layout scannable, a critical aspect of web design. Layouts with well-managed white space allow users to scan information and groups of information to find what they're looking for quickly. Cluttered layouts, or ones that don't effectively manage white space, make it hard for the user to identify patterns that are essential for scanning information. Imagine a group of people milling around at a party versus a line of soldiers at roll call. The people are the same, but the space between them has been organized.

This is a side-by-side comparison of a competition mini-site created by the AIGA DC. On the left is the original site; on the right the white space has been filled in to highlight the consistent and almost rhythmical use of space. The generous spacing around the headline and lead-in statement helps them stand out on the page. The non-default, slightly open line spacing for all the text gives the pages a very light and scannable feel.

White space is actually a reference to "ground" as in "figure-ground," and doesn't need to be white at all. In this example, ThinkingForALiving.com, the ground is a pink hue, but the result of well-constructed white space on the design is the same.

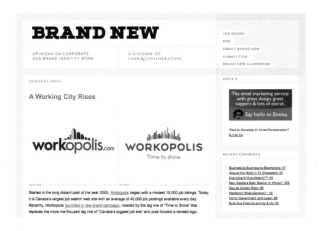

Containment

At times, more than space is needed to highlight, group, or separate elements on a page. Borders, lines, and boxes can be helpful in defining the space and containing elements within subgroups. The varying types of borders that can be created with CSS, including dotted, dashed, double, and single lines, make them powerful stylistic elements as well. Even rounded corners, a popular design treatment for boxes, are now possible using CSS3, and they are viewable in browsers compatible with CSS3.

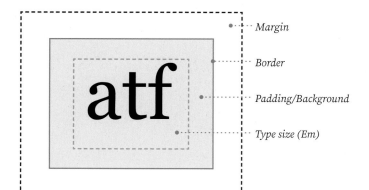

The sites seen here, BrandNew. UnderConsideration.com (top) and 20x200. com (bottom), use a wide variety of distinctive line styles to segment the page and reinforce a design style.

(Left) CSS can be used to define the border of an object. The border, represented by the orange line in this diagram, lies between the padding distance and the margin area.

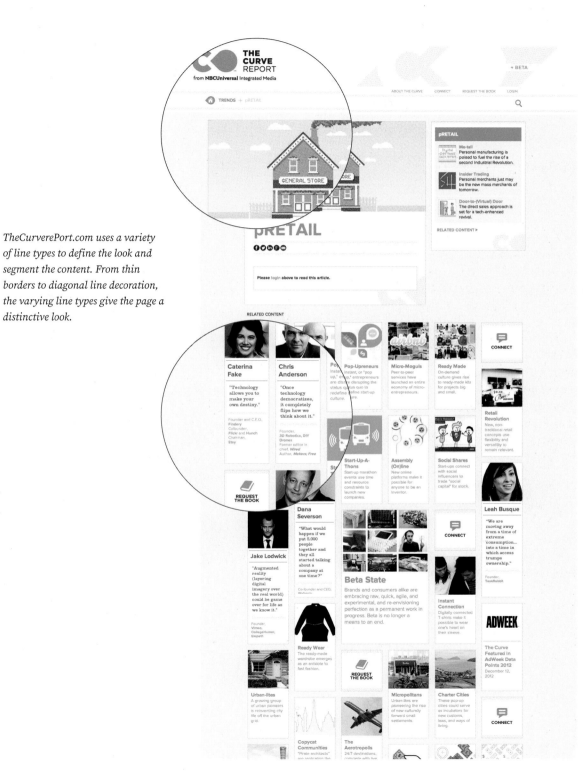

TheCurverePort.com uses a variety of line types to define the look and segment the content. From thin borders to diagonal line decoration, the varying line types give the page a distinctive look.

Grids not only organize the elements of a design, they organize the **space** within a design.

Grids

One of the oldest ways to create a balance of figure and ground is through the use of a grid system. Grids not only organize the elements of a design, they organize the space within a design. Clearly aligning design elements through the use of columns creates defined space, and it's this space that gives the appearance of organization.

Grids are made up of columns (where the content goes), gutters (the space between columns and margins), and the space around the perimeter of the layout. By carefully defining these attributes, web designers can control the density of a page, which is the amount of detail that is visible to the user. Dense pages tend to be harder to read, although pages that lack sufficient density can appear un-unified and fall apart.

Early websites were laid out using tables, a word processing convention of rows and columns used to arrange elements. Some early web layouts had a compartmentalized or checkerboard feel as a result of using or overusing tables. Tables are also limited in their flexibility and result in long markup for even simple layouts. Although tables still exist in HTML, <div> or divider tags have taken over as the preferred method of containing and laying out elements of a design. The flexibility of CSS-styled <div> tags more closely resembles the feel of a print layout program such as Adobe InDesign. They enable very sophisticated print-like layout and grid use.

New Graphic Design *magazine was started in 1958 by Richard Paul Lahose, Josef Müller-Bockmann, Hans Neuburg, and Carlo Vivarelle. The cover of issue 16, pictured on the opposite page, illustrates the grid system that permeated the entire magazine and is credited with defining the Swiss style of graphic design.*

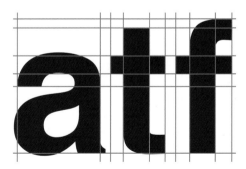

The letterforms of Helvetica, the ubiquitous Swiss typeface and subject of a documentary film, are based on a grid system, making it instantly recognizable over its predecessor, Akzidenz Grotesk.

Neue Grafik
New Graphic Design
Graphisme actuel

Internationale Zeitschrift für Grafik
und verwandte Gebiete
Text dreisprachig
(deutsch, englisch, französisch)

International Review of Graphic
Design and related subjects
Issued in German, English and French

Revue internationale du graphisme et
des domaines annexes
Parution en langue allemande,
anglaise et française

16

Herausgeber und Redaktion
Editors and Managing Editors
Editeurs et rédaction

Druck Verlag
Printing/Publishing
Imprimerie Edition

Richard P. Lohse SWB/VSG, Zürich
J. Müller-Brockmann SWB/VSG, Zürich
Hans Neuburg SWB/VSG, Zürich
Carlo L. Vivarelli SWB/VSG, Zürich

Walter-Verlag AG, Olten
Schweiz Switzerland/Suisse

Pattern Library
Grid System

Grid System

Typography

Form Elements

Navigation

Tables

Lists

Slats

Stats/Data

Feedback

Grid sizes   Grid gutter   Mixed grids   Responsive columns   Grid example

Our grid system is composed of 8 flexible columns with a gutter between columns of 30px. We apply border-box so that the border and padding is included in the width of the grid columns.

Grid Sizes

Size 1 of 1

Size 1 of 2

Size 1of 3

Size 1of 4

Size 1of 8

```
1   <div class="line">
2     <div class="unit sizelof3">
3     </div>
4     <div class="unit sizelof3">
5     </div>
6     <div class="lastUnit sizelof3">
7     </div>
8   </div>
9
```

Notes

When using the grid, wrap the columns using a line and use lastUnit for the last column. Refer to OOCSS base classes to learn more about the grid classes.

This example applies to the other ratios we support: 1/1, 1/2, 1/3, 1/4, and 1/8

Grid gutter

Our grid columns have a 15px padding on either side that results in a 30px gutter between columns and a 15px gutter on the grid edges. Even though our columns are fluid, the gutter remains constant.

Mixed Grids

The grid layout is easily extended by nesting and mixing different column sizes.

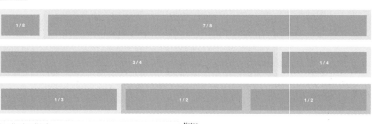

```
1    <div class="line">
2      <div class="unit sizelof3">
3      </div>
4      <div class="group sizelof3">
5        <div class="unit sizelof2">
6        </div>
7        <div class="lastUnit sizelof2">
8        </div>
9      </div>
10   </div>
11
```

Notes

When nesting grids use the group class to eliminate gutters on the parent element of the columns that are being nested.

Wherever possible, limit nesting grids to 2 levels deep. Anything more than that would require the use of nopadding on grid columns or units for correct grid alignment and indentation.

90

The design team at MailChimp.com gives users a glimpse at their design system with their Pattern Library. This tool reveals the systems, including this grid system behind the site.

CRW / CORPORATE RISK WATCH

| Profile | Services | Case Studies | Regions | Contact |
| who we are | what we can do | problem solutions | where we operate | enquire here |

Due Diligence

An international oil company was considering entering into a business relationship with an oil and gas producer in the Philippines but suspected that the target company was associated with local politically exposed persons and that this association might have favoured the company in obtaining a concession for oil extraction. To comply with the Foreign Corruption Practices Act regulations it was necessary to conduct an extensive due diligence to assess the potential risks attached to the deal.

A systematic analysis of publicly available documentation in the Philippines and discreet source enquiries into the target company and its principals were conducted.

It emerged that the management of the oil producer in the Philippines was composed of highly experienced and prominent figures from the public energy sector who continued to retain significant political influence. The beneficial owner of the company in the Philippines was hiding behind nominees and offshore structures but his identity was revealed through discreet enquiries with sources in the local energy sector. It emerged that the ultimate beneficial owner was a former representative of the local government and that his political influence enabled the company to obtain the said concession. The risks attached to the target company were assessed.

Competitor Intelligence

A British company operating in the IT sector was interested in the purchase of one of its three Italian competitors but was unable to put in place the right strategy without having an in-depth knowledge of the Italian IT sector and specifically, the three target companies. In addition the client suspected that one of the players had links to the Organised Crime but was unable to assess the veracity of this rumour.

The work conducted included analysis of the financial situation, business models, investments, marketing and product strategies with respect to each of the three companies through a systematic retrieval, analysis and cross examination of publicly available information, combined with discreet source enquiries with local industry experts.

The work resulted in the identification of one of the tree competitors as the potential acquisition target. Evidence was obtained confirming the allegation of association with organised crime by one of the target companies.

Litigation Support

A Dutch operator in the printing sector suspected that a former employee, an engineer who had worked for the company for over twenty years and who had recently retired, was providing a competitor with the company's know how and other confidential data such as supplier and client contacts. To get these activities to stop, the Dutch operator initiated a legal proceedings against the competitor and its former employee but did not have sufficient evidence to prove the case.

The work conducted consisted in collecting evidence, both factual and testimonial in support to the client's claim, including surveillance and witness identification.

The client was able prove with factual evidence the case of unfair competition. The competitor stopped to act unfairly and the client received compensation for the damages suffered.

Left Loft, the designers of CorporateRiskWatch.com, actually expose the grid structure they're using by tracing it with dotted lines. The elements of every page seem to dance around this five-column grid.

CRW / CORPORATE RISK WATCH

| Profile | Services | Case Studies | Regions | Contact |
| who we are | what we can do | problem solutions | where we operate | enquire here |

CRW relies on a multi-lingual team with international experience in risk management and on a network of contracted professionals worldwide.

The success of our clients' businesses is influenced by decisions taken with respect to new partnerships, investments and business dealings.

CRW's skilled team of multi-lingual professionals with international experience helps clients mitigate the exposure to financial and reputational risks.

CRW provides clients with reliable information and strategic analysis they require to maximise business opportunities in different regions of the world.

CRW offers services to comply with anti-corruption and anti-money laundering legislations and in support of business partnerships, investments and market entries, hiring of employees, complicated business transactions and legal disputes.

Memberships

Corporate Risk Watch is the holder of a private investigations license, in accordance with the paragraph 134 T.U.L.P.S. issued by the Italian authorities.

Corporate Risk Watch is a member of the following associations:

Association of anti-money Laundering Specialists (www.acams.org); Italian-American Chamber of Commerce in Italy (www.amcham.it); Italian-Chinese Chamber of Commerce in Italy (www.china-italy.it)

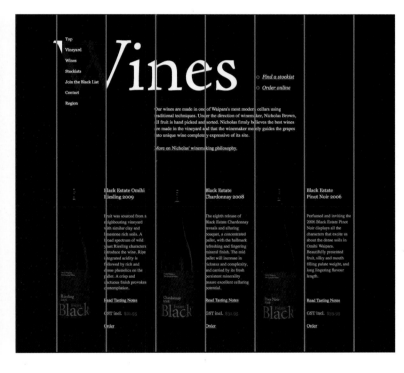

BlackEstate.co.nz, which has won numerous awards for its use of typography and unique navigation, features a six-column grid. The tall page is held together because of the strict adherence to the elegant grid.

The Swiss styling of WilsonMinor.com is a classic example of a well-used grid structure. Headlines, subheads, images, and text work together to define and span the six-column grid.

The grid on DigitalPodge.com is filled in a more organic way. Instead of the elements neatly aligning in exactly the same way, there's a playful bouncing of text and image within the grid structure.

■NY

Championing the future of design for all.

Mission
Leadership
Staff
Collaborators
History
Sponsors

Leadership

Staff

Collaborators

History

Sponsors

Made in NY

Once the grid system has been established, elements of the design are placed within the grid. Objects can span more than one column width, but each element must have some clear relationship to the grid itself. Any element that relates to the grid in a unique way or breaks the grid system will rank higher on the hierarchy scale.

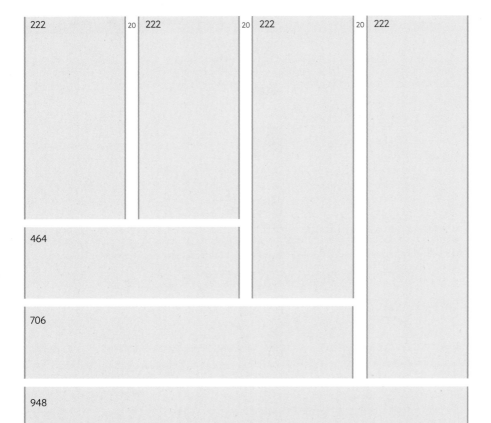

This is a diagram of a grid system with the following specifications:
Width: 948 pixels (px)
Columns: 4
Column width: 222px
Gutter width: 20px
2-col span: 464px
3-col span: 706px

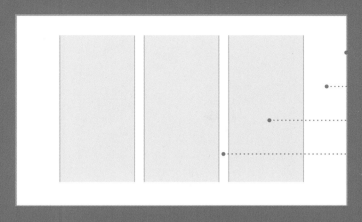

- Browser frame
- Margin
- Column (3-column grid)
- Gutter

FIXED WIDTH (FLOATING CENTERED; FIXED LEFT)

Grids used for web design need to have a flexible quality to them in order to accommodate varying monitor widths and resolutions. There are several solutions to this issue. In this example of a fixed-width grid, the grid either floats in the center of the browser window or is fixed to the left side. As the browser window expands in both cases, the layout within the grid is not altered.

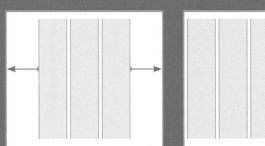

VARIABLE WIDTH

In a variable-width grid system, each column expands proportionately with the width of the browser frame. This causes the layout within the grid to change and shift depending on the width of the user's monitor.

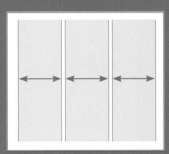

COMBINATION OF VARIABLE AND FIXED WIDTH

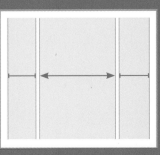

This diagram shows a grid that has both fixed-width columns as well as a single variable-width column. As the browser window expands, only one column width expands with it. The layout of the center column shifts, while the two flanking columns stay fixed.

The grid system on SimpleArt.com. au is flexible, so whether the page is viewed on large or small monitors the layout feels consistent. Note in the wider layout below, the columns of the grid widen and the header/navigation area moves to the right.

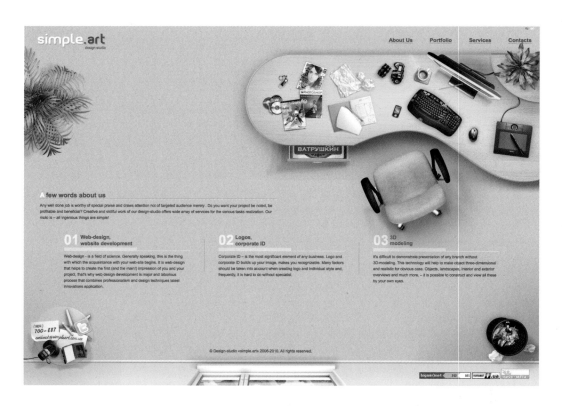

Items in a layout that break the established system stand apart from the rest of the elements within the system. In this example of AIGALosAngeles.org, the AIGA logo does not "sit" on the grid. By shifting outside the grid it's given more visual value than the other elements on the page, as illustrated in this diagram.

The Baseline Grid

Something that print designers have been using for years but is only recently being adopted by web designers is the use of a baseline grid. A baseline grid is a horizontal grid system that exactly aligns the baselines of all the text on a page, regardless of size or style. Baseline grids create a smooth rhythm in the typography within a design.

TheGridSystem.org is a blog about the use of grids in design. An interesting feature of this site is the ability to expose the grid structure as well as the baseline grid.

Principles of Web Design

Creating a baseline grid in CSS involves a bit of math, since there's no built-in baseline grid attribute. A web designer starts by choosing a type size for the majority of the text on the page. Then a line height is applied in the CSS, which is essentially the equivalent of leading. To create the appearance of a baseline grid, all other measurements, including the margin spacing, display type size, etc., should be multiples of the line height. This will ensure that all baselines will line up relative to one another.

This is an example of a baseline grid in use. Note that each typographic element, regardless of size or typeface, sits exactly on the baseline grid.

From Tattly to Time *magazine, users are browsing the web on mobile first, studies show. These two sites are excellent examples of sites that provide a unique experience on the narrow screen of a phone.*

Responsive Design

This chapter is about space. Responsive design is about reorganizing space to maximize the visibility of key design elements on varying screen sizes. Responsive design is an evolution of web layout because of the prevalence of people browsing web pages on mobile devices. Design considerations for small screens is vastly different than for large desktop screens, hence the need for a design to respond/change based on the size of the screen on which it is being viewed.

Responsive design is much more than the simple rearrangement or stacking of content "blocks" so that they run vertically down a slender mobile screen. Great responsive design also takes into account things like page load times and typographic legibility to completely change the content and design of a page for mobile devices. With responsive design a designer can hide large imagery on a mobile device, or change the color and size of type to increase the contrast and legibility.

Without getting too technical, responsive design works through what are called CSS (cascading style sheet) media queries. These media queries read a user's browser data to determine the width of the browser window before loading the styling for a page. Currently, there are three primary break points that designers use for their media queries: larger than 768 pixels wide (desktop), less than 768 pixels (tablet), and less than 480 pixels wide (mobile device).

TIME (desktop view)

TIME

f y g+ t 🔊 Apps

Search TIME

Home | NewsFeed | U.S. | Politics | World | Business | Tech | Health | Science | Entertainment | Video | TIME 100 | Photos

Magazine | LIFE | Opinion | Weather

Thursday, February 20, 2014

■ Breaking Al-Jazeera journalists trial starts in Egypt

Bulent Kilic / AFP / Getty Images

Allies Abandon Ukraine Leader

By Simon Shuster / Kiev

The nation's embattled President finds his inner circle becoming increasingly smaller as the revolution's death toll climbs

- E.U. Imposes Sanctions on Ukrainian Officials
- 📷 Battleground Kiev: 'Halfway Between a War and A Protest'
- Ukraine Inches Ever Closer to a Full-Blown Civil War
- Ukrainian Skier Withdraws From Olympics to Protest at Home

2 Americans Found Dead on *Captain Phillips* Ship

Pussy Riot Releases Music Video of Sochi Beating

VIDEO

Kiev's Frontline From Instagram: A View From The Ground

00:00 — 01:41

Embed Email Share

Kiev's Frontline From Instagram: A View From The Ground

'The Lego Movie' Animators Show How the Film Was Made

Korean Families Reunite After 60 Years of Separation

TIME Subscribe Now ▶

POLITICS
Hollywood Is Coming to Capitol Hill

Obama and Canadian Leader Make Beer Bet on Olympic Women's Hockey Final

George Bush Misses Air Force One

WORLD
E.U. Imposes Sanctions on Ukrainian Officials

Al-Jazeera Journalists Trial Starts in Egypt

Pussy Riot's New Music Video Shows Sochi Beatdown

BUSINESS
Samsung's Ruthless New Ad Mocks Apple's Latest

Target Shoppers Shrug Off Massive Credit Card Data Breach

The Surprising Best Thing about Google Fiber Coming to Your Town

HEALTH
Is Too Much Tanning a Mental Illness?

4 Diet Secrets of the U.S. Olympics Women's Hockey Team

The Internet is a Safer Place for Your Teen Than You Think

TECH
Facebook's WhatsApp Acquisition Explained

Microsoft Stops Hiding Office's Free Online Edition

The Thief Launch Trailer Has Everything, Including Electric Guitars

ENTERTAINMENT
Beloved TV Show *Pushing Daisies* May Become A Broadway Musical

Lorde's Collaboration With Disclosure Has Given Us the Best 'Royals' Remix Yet

Kanye West, Elton John, Skrillex to Perform at Bonnaroo 2014 (Probably Not All At Once)

PHOTOGRAPHY

MAGAZINE

🔒 Empty Slopes

🔒 Joel Stein: DNA of Champions

🔒 The Tonight Show's Upworthy New Host

🔒 Frozen's Hot Following

The Magazine
Subscribe

The Strange World of Airline Cancellations

Party Foul in U.S. Politics

A Step Backward for Labor

Young Kids, Old Bodies

Table of Contents
Subscribe Now
Online Issue Archive

Facebook Snags WhatsApp In Its Biggest Buy Yet
Social networking giant purchases the global messaging app for $19 billion

- Facebook Rejected WhatsApp Founder For a Job in 2009
- Report: Google Also Wanted WhatsApp With $10 Billion Bid

Fire Forces Evacuation of Iowa Town
Responders battle blaze at nearby fertilizer plant

Kansas Spanking Bill Gets Spanked
It would've permitted up to 10 strikes on a child's behind

Obama, Canadian Prime Minister In Beer Bet
The real question is what kind of brew is at stake

CEO Gives Harvard $150 Million
Hedge-fund head offers largest gift ever

Alaska's Road From Nowhere
Government rejects gravel path, leading to deadly problem

Gravity Won't Win Best Picture
Corliss on the cinematic glory that will doom it

New *Fantastic Four* Finds Its Superheroes
But none of the stars are household names

Latest Headlines

- Pope Francis Meets With Prisoners at His Home
- 27 Liberal Groups Oppose Obama Judicial Nominee
- ● One Californian Won the $425 Million Powerball
- Bitcoin ATM Arrives at Boston Rail Hub
- Natural Gas Prices Surge to a 5-Year High
- Miami Dolphins Fire Coach After Harassment Report
- Idaho Couples Sue to Overturn Gay Marriage Ban
- 4 Accused of Cutting Swastika into Classmate's Head
- ● Tearful Korean Reunions Begin; First Since 2010
- ● CNN: Where the Middle Class Thrives

Editor's Picks On Weather.Com

London's River Thames Breaches Banks

America's Best Winter Drives

Winter Storm Pax Forecast: Southern Snow and Ice

Most Read | **Most Emailed**

1 Russian Figure Skater Takes a Fall, and More Surprises from the Ladies Short Program

2 Tales from the TSA: Confiscating Aluminum Foil and Watching Out for Solar Powered Bombs

3 Soda Wars Bubble Up Across the Country

4 Alaskan Outrage As Obama Appointee Rejects Wilderness Road

5 Two Americans Found Dead on 'Captain Phillips' Ship

6 The One Word You Need to Stop Using Immediately

7 Britney Spears Forgot to Lip Sync During Her Las Vegas Show. But That's Okay

8 Now There's Another Reason Sitting Will Kill You

9 The New *Fantastic Four* Has Found Its Superheroes

10 Here's Why Your Netflix Is Slowing Down

Person Of The Year

The People's Pope
He took the name of a humble saint and then called for a church of healing. Read more about Pope Francis and see the rest of the shortlist.

TIME (mobile view)

TIME Follow Apps

SECTIONS ▶

■ Al-Jazeera journalists trial starts in Eg...

Bulent Kilic / AFP / Getty Images

Allies Abandon Ukraine Leader
By Simon Shuster / Kiev

The nation's embattled President finds his inner circle becoming increasingly smaller as the revolution's death toll climbs

- E.U. Imposes Sanctions on Ukrainian Officials
- 📷 Battleground Kiev: 'Halfway Between a War and A Protest'
- Ukraine Inches Ever Closer to a Full-Blown Civil War
- Ukrainian Skier Withdraws From Olympics to Protest at Home

Facebook Snags WhatsApp In Its Biggest Buy Yet

- Facebook Rejected WhatsApp Founder For a Job in 2009
- Report: Google Also Wanted WhatsApp With $10 Billion Bid

2 Americans Found Dead on *Captain Phillips* Ship

Pussy Riot Releases Music Video of Sochi Beating

DON'T MISS

Fire Forces Evacuation of Iowa Town
Responders battle blaze at nearby fertilizer plant

Kansas Spanking Bill Gets Spanked
It would've permitted up to 10 strikes on a child's behind

Obama, Canadian Prime Minister In Beer Bet
The real question is what kind of brew is at stake

CEO Gives Harvard $150 Million
Hedge-fund head offers largest gift ever

Alaska's Road From Nowhere
Government rejects gravel path, leading to deadly problem

Gravity Won't Win Best Picture
Corliss on the cinematic glory that will doom it

New *Fantastic Four* Finds Its Superheroes
But none of the stars are household names

VIDEO

● Kiev's Frontline From Instagram: A View From The Ground

● 'The Lego Movie' Animators Show How the Film Was Made

● Korean Families Reunite After 60 Years of Separation

Responsive design is **much more** than the simple rearrangement or stacking of content "blocks."

Therefore, designers need to create three individual, yet related, grid systems for a single page layout.

For desktop (over 768 pixels wide), a common grid for design purposes is sized between 950 pixels and 990 pixels wide, but it can be up to 1,200 pixels as monitor resolutions continue to increase. Once the width has been determined, a designer decides how many columns are needed. More columns means more design flexibility; however, too many columns can make recognizing relationships difficult. There is no right number of columns, but the optimal grid gives a layout a clear sense of organization while still allowing for flexibility. The column width for a grid is determined by the overall width divided by the number of columns. And finally, gutters, or the spaces between the columns, are added, providing separation between the elements in each column.

A tablet layout (less than 768 but more than 480) shares many relationships to the desktop grid. Generally, the number of columns would be reduced by half and the gutter widths decrease slightly. Often, designers will remove the margin, or space surrounding the page, including a background image or pattern, to maximize the useable space.

Finally, a mobile grid (less than 480 pixels wide) is reduced to a single column. Given the narrowness of the screen, more than one—or possibly two—columns causes issues of legibility and usability as design elements, including buttons, get smaller. Often the navigation changes to a drop-down menu and much of the imagery is removed from the page by the CSS to conserve download times.

Desktop > 768 pixels wide

Tablet < 768 pixels but > 480 pixels wide

Mobile < 480 pixels

Responsive design **fluidly responds** to the width of a browser. Adaptive design generally has two to four **pre-formatted design states**.

Responsive vs. Adaptive Layouts

Often mistakenly used interchangeably, responsive and adaptive design are slightly different. Responsive design fluidly responds to the width of a browser, forming a clear layout at any width between 480 pixels and more than 768 pixels. Adaptive design generally has two to four pre-formatted design states that it "snaps" to depending on the width of the browser. This offers a designer a bit more control over the layout as there are no in-between sizes that can sometimes produce visually awkward layouts. While it is more common to produce a responsive design, adaptive design can be very useful, especially if the target user group is small and its technology is well defined.

(Opposite) Carters.com, the children's clothing store, takes a unique approach to their mobile site. Much of the content from the desktop site is stripped away in lieu of navigation. This is done for two reasons: to help with the speed of the download of each page (saving data charges for the user) and to expedite the shopping process.

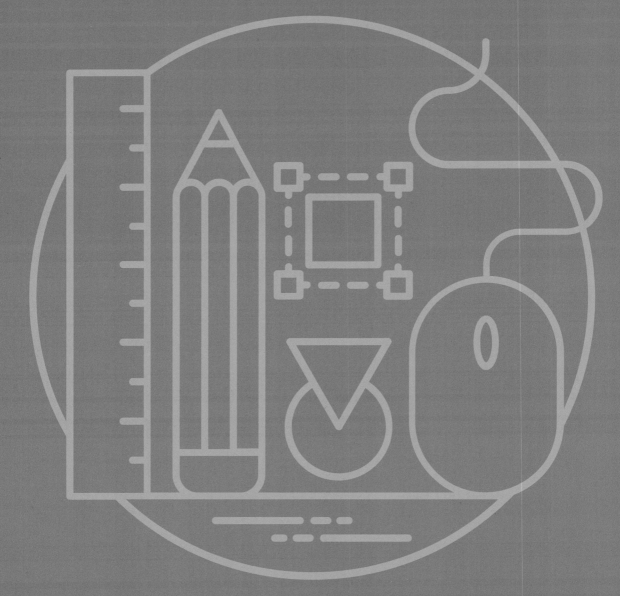

Section II

Design

CHAPTER 4

Anatomy of a **Web Page**

1. Header
2. Navigation
3. Feature
4. Body/Content
5. Sidebar
6. Footer
7. Background

Form and Function of Web Design

Web design, like any other form of design, requires the designer to understand the end user's habits, the context in which the work is received, and the necessary function of the end product. These factors usually present limitations that set the boundaries for starting a design project. For web design, these boundaries have caused several design and structural conventions to emerge. Such conventions include a page header, persistent navigation, content areas and sidebars, footer navigation, and often a background treatment. Although styling and aesthetics vary greatly from site to site, most sites adhere to this basic structure. Each of these common web design elements, and their placement on the page, came to be for several basic reasons.

THE NATURE OF HOW THE PAGES ARE VIEWED
In Western culture, we're conditioned to read from left to right, top to bottom. Therefore, the natural position for important information would be the upper left of a web page. This ensures that elements such as logos, navigation, and "featured items" are perceived first by the user.

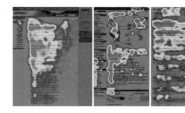

These images show the results of an eye-tracking study. They indicate that users focus their attention on the upper-left area of a web page.

The notion that users scan pages from left to right, top to bottom, has been validated through the use of eye-tracking studies. Sophisticated cameras fixed to the top of a computer screen have the ability to track the eye movements of internet users and map out the patterns. The red areas in the images to the left indicate where users focused most of their attention. They reveal not only the fact that users' attention is mainly focused on the upper left of a page, but also that web users skim a page for key points, as shown by the spotty bits of color in the center and left images.

Many web design conventions are borrowed from the world of print communication. Pictured here is the New York Times newspaper showing a header and feature area very similar to those on a web page.

The "fold"

BORROWED CONVENTIONS

Because almost all early web designers were amateur designers or trained as print designers, elements from print design were converted to web design. Design elements like headers, feature areas, body text, and sidebars all come directly from age-old newspaper design standards.

The "fold" of a newspaper is literally the horizontal crease in the center of the front page delineating the top half from the bottom half. Newspaper editors tend to put as much of the most important information as possible above that fold since that's the area that potential newspaper buyers will see. Similarly, a "fold" on a web page is the line that delineates where the browser window cuts off the content. Areas above the fold are seen by the user when the page loads. Content below the fold requires that users scroll down.

USER EXPECTATIONS

Sites that want to attract the masses, like news portals, travel sites, e-commerce sites, etc., need to appeal to the lowest common denominator in terms of one's ability to use technology. As the web became established in the mid- to late 1990s, companies interested in having their users find what they wanted quickly would imitate the metaphors for navigation and site layout from other, already established, sites. For example, Amazon.com is credited with creating the first tab-style navigation (another borrowed convention); although there are probably earlier examples, the "tabs" served as a metaphor that worked in part because tabs were something people understood from the "real world" of file folders. As a result, websites all over the internet began using a tab structure for their navigation—and still do to this day. Even Apple.com, known widely for its innovative design, once used a tabbed navigation very similar to that of Amazon.

Many web design conventions like the tabs seen above are abstracted versions of real-life objects.

At the height of the tab craze in 2000, some said that the navigation on Amazon.com resembled a graveyard.

Google

Google Search   I'm Feeling Lucky

SEARCH ENGINE OPTIMIZATION (SEO)

Having a high search engine rank is critical to a company's online success. A higher rank on a list of search results means more traffic. Search engines, such as Google.com and Bing.com, use various methods to evaluate the content of a site and determine its rank. Some design factors that influence the search engine optimization of a page include: text links in the main navigation; multiple keyword-rich text links throughout the page; limited use of images, especially images of text, since search engines cannot get content from images; bolded subhead copy styled with the <H> tags; and important content placed above the fold—the higher the better. Although these are not all of the SEO factors that influence the rank of a page, these are generally the factors that a designer has the most control over. The topic of SEO is discussed further in chapter 7.

Orbitz.com is a good example of a page designed for SEO. Multiple keyword-rich text links, bolded subheads, and limited use of imagery consistently produce a top ranking for searches of "Vacation Packages."

ADVERTISING STANDARDS

The Interactive Advertising Bureau (IAB.
net) was established in 1996 to set up
standard practices in web advertising.
The organization sets forth rules that
govern the size, shape, and file weight
(among other things) for advertising
assets. This helps advertisers create
a finite series of banners that can be
used on any website that adopts the IAB
standards. For web designers, this means
that their web design must accommodate
banners that are 300 x 250 pixels ("big
box"), 180 x 600 pixels ("skyscraper"),
and/or 728 x 90 pixels ("leaderboard"),
among others. If a website is funded with
ad revenue, these dimensions become
a critical part of the framework of the
site. Additionally, advertisers want their
ads above the fold so that the user sees
them immediately. Website owners,
on the other hand, don't want the ads
to overpower the message of the site.
Web designers satisfy both sides by
establishing a structure that flows with
the required sizes of the ads—a 300-pixel-
wide sidebar will fit a big box ad without
any dead space around it, for example.

*Time.com and many other sites across the web
display advertising. In this example of the home
page, a leaderboard ad appears in the header.*

Without **understanding the function** behind standard web design conventions, designers are purely imitating things that they've seen.

While these particular factors are unique to web design, the idea of a set of parameters that restrict and inform a design is not unique. Car designers, for example, are faced with hundreds, if not thousands, of these types of challenges. People want to be able to drive more than one make of car without having to work to relocate and decipher the speedometer, for instance. Yet, there's a wide range of variation in the sizes and shapes of cars on the road today.

The duality of form and function is a universal design concept; however, most new web designers aren't as aware as they should be of the technical and functional implications behind the design decisions they make. Without understanding the function behind standard web design conventions, designers are purely imitating things that they've seen. This chapter explores the parts of a web page and specifically how those parts contribute to the overall effectiveness of a site—aesthetically and technically.

Car designers face similar challenges as web designers when designing a dashboard interface. They seek a balance between unique style and standardization and ease of use.

There's a template for that . . .

For better or worse, the web is becoming increasingly templat-ized. Services like Squarespace, Wix, Shopify, Wordpress, and Adobe Portfolio, just to name a few, offer a means for content producers to quickly select a template and arrange their content in standardized layouts. So where does the designer fit in to this? It's a good question that is becoming more and more difficult to answer. However, there will always be a need for design professionals who understand hierarchy and are sensitive to the nuances of screen-based design like typography color and space. The examples below show how a single template can be used for multiple types of content.

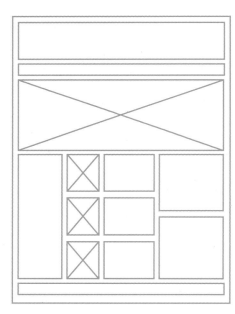

MarthaStewart.com is an elegant design
example from both a structural as well as
an aesthetic point of view. The subtle and
consistent design treatments give the site a
uniquely Martha Stewart feel, despite using
a standard web structure.

martha stewart

🔍   Martha's Blog   Your Account ▾   Login   Sweepstakes

☰ EXPLORE   FOOD   HOLIDAYS   ENTERTAINING   HOME   GARDENING   CLEANING & ORGANIZING   WEDDINGS   SHOP   SUBSCRIBE

What's New

The 10 Best Sheet Sets to Help You Get a Better Night's Sleep, Based on Customer Reviews

How to Attract Owls to Your Backyard

The Martha Blog
Martha Shares an Up Close and Personal Perspective of Her Life

A Cook's Guide to Choosing and Preparing Leeks

Martha's Five Rules for Raising Pets That Live Together Harmoniously

"The key to living harmoniously with pets is taking the time to train, nurture, and care well for them," she says.

Get Inspired

Get Ready for Barbecue Season with the Help of These Two New Cookbooks

The Glow Pro: Rose-Marie Swift, the Founder of RMS Beauty, Outlines Her Recipe for Clean Living

Set the Perfect Spring Table with These Essential Martha Stewart Products

Your Family Will Love These Grilled Buttermilk Chicken Tenders with Dipping Sauces

Crusts and Toppings Galore: Our Best Pizza Recipes

Fresh Ideas

Three Instant Ways to Digitize All of Your Photographs

Classic Recipes

No-Bake Cheesecake

Basic Pancakes

Perfect Hard-Boiled Eggs

Simple Crepes

Corned Beef and Cabbage

Perfect White Rice

Why Do Eyebrows Thin Over Time?

The header graphic for RolllingStone. com uses the magazine's iconic logo as the central element. The clean, centered design approach creates a unique and identifiable presence for the brand.

Header

The header of a web page is one area that remains relatively consistent throughout a website. It acts as a grounding force for the user by identifying and visually unifying all the pages of a site. Headers establish the brand look and feel for a site and often will present the user with a call to action—search, buy, register, etc. The header of a page must perform these tasks without overpowering the content of the page and distracting the user.

Because the header area tends to stay consistent from page to page, it is often where the client's logo appears. It has become a common expectation of users that the logo on a site, specifically one located in the header of a page, will link the user back to the home page.

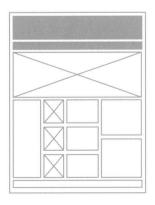

Headers act as a grounding force for the user by **identifying** and **visually unifying** all the pages of a site.

The code behind the header contains information that is vital to the search engine optimization of the page. From metadata (keywords and descriptions of the page in the code) to the page title (this is the line of copy that appears on the top of a browser window), search engines use these elements to begin indexing the content of the page.

The header on 99u.com is a fixed, or "sticky," header that does not scroll with the page. In its initial state, the header takes up a good amount of space to accommodate the logo, tagline, social links, and a search feature. As the page scrolls (bottom), the navigation slides over, and the logo appears in the blue bar. This is useful for long pages as it gives users access to the navigation, even in the middle of the page.

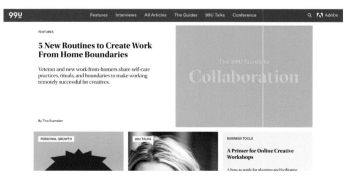

Newsweek.com's bold use of their brand color and centered placement of their logo make for a distinctive page header.

Staples.com is an example of an extremely functional header. The logo is flanked by shopping tools like the search bar and the shopping cart.

Mon, May 17, 2021

Newsweek

LOGIN   **SUBSCRIBE ›**

U.S.   World   Business   Tech & Science   Culture   Newsgeek   Sports   Health   Opinion   Experts   Vantage   Search

TRENDING   JOE BIDEN   DONALD TRUMP   CORONAVIRUS   VACCINE   GAVIN NEWSOM

FEATURED STORIES

POLITICS

Liz Cheney Likens Trump Election Conspiracies to Chinese Communist Party's Attacks on Democracy

Cheney vowed to "restore" the Republican Party's character in the wake of Trump's continued election conspiracies and fanatical supporters' lingering in the GOP.

Majority of Republicans Agree With Cheney Ouster, Say She's Not 'on Message' With GOP: Poll

TECH & SCIENCE

Elon Musk Confirms Tesla Has Not Sold Bitcoin After Price Crash

The Tesla CEO's tweet followed speculation that his company was planning on dumping holdings of the cryptocurrency.

U.S.

Michigan Newspaper Calls Gov. Whitmer 'Clueless' Amid Pipeline Dispute

The Detroit News said the Democratic state leader had proposed "ridiculously inadequate" alternatives to the pipeline.

U.S.

UFO Capabilities, 'Compelling' Evidence Revealed by Former Pentagon Expert

A report by the Director of National Intelligence and the Secretary of Defense on UFO analysis is due to be released by next month.

May 17 Tax Deadline Explained: What Time Are Taxes Due?

The deadline for filing individual tax returns in the U.S. was pushed back from the usual date of April 15 to May 17 this year.

SPONSORED INSIGHT

LensCrafters: Explore the New Eyewear Collections

The world's top designers have unveiled new eyewear for spring and summer. Make a bold statement with cool frames from brands like Ray-Ban, Versace, and Prada, tailored to fit your vision needs and lifestyle.

TOP STORY

NEWS

Exclusive: Inside the Military's Secret Undercover Army

Thousands of soldiers, civilians and contractors operate under false names, on the ground and in cyberspace. A Newsweek investigation of the ever-growing and unregulated world of "signature reduction."

MY TURN

'I Help Polyamorous Couples With Relationship Problems'

'I'm 13, I Have an IQ To Rival Einstein'

'I Helped Dying COVID Patients Say Goodbye. I Can't Forget'

CULTURE & TRAVEL

CHRISTIE'S   TRAVEL   FILM

Wine That Made Trip to Space on Sale With $1 Million Price Tag

These Sustainability Innovations Around the World are Revolutionary

The 'Avatar' Connection to Cameron's Whales Doc: They're Just Like Us

MORE STORIES

Ex-Officer Kim Potter To Appear in Court Over Daunte Wright's Death

The city's police chief, who resigned after Wright's shooting, claimed Potter shot Wright by accident.

5 Past UFO Sightings As UAP Report to Be Released by Task Force

The report follows the official release in 2020 by the Department of Defense of three videos showing Unidentified Aerial Phenomena.

How Hard Gas Shortages Are Hitting Each State, According to Drivers

New data shows most gas stations in D.C. and North Carolina faced gas outages on Sunday evening.

Walmart Offers Workers $75 Incentive to Get COVID Vaccine

The company also said fully vaccinated employees could work without a mask starting from May 18.

Prince Harry Damaged Loving, Caring, Fun Relationship With Charles: Butler

Prince Harry's former butler told a documentary the duke may not ever repair his relationship with father Prince Charles after a slew of public criticisms.

Watch Video of NASA Rocket Launch From Wallops Island After Days of Delays

The Black Brant XII rocket took off as part of the KiNET-X mission to investigate how energy from the sun interacts with the Earth's magnetic field.

CNN Drops Freelancer Adeel Raja Over Pro-Hitler, Anti-Semitic Tweets

The network said it would not work with Raja again "in any capacity" after his anti-Semitic tweets were surfaced.

The Full Story Behind Prince Harry's Comments Criticizing Royal Upbringings

Prince Harry says he suffered from "genetic pain" passed on by his father in comments that echo a bombshell biography criticizing the royal family—produced with Prince Charles' help in 1994.

3 States With Anti-Trans Laws To Host Softball Regionals Despite Threat

It's a stark contrast from the NCAA's stance last month when it said it would be difficult to hold championship events in states that have anti-transgender sports laws.

Netanyahu Says Journalists in al-Jalaa Tower Weren't in Danger

"One of the AP journalists said, 'We were lucky to get out.' No you weren't lucky to get out. It wasn't luck." Israel's prime minister said of the airstrike.

Marjorie Taylor Greene Fires Back at Rashida Tlaib Over Israeli Airstrike

"The #JihadSquad supports re-entering the Iran deal &

THE DEBATE

Ban Critical Race Theory Now
BY MAX EDEN

VS

The Republican Push to Ban Critical Race Theory Reveals an Ugly Truth
BY MARCUS JOHNSON

The Debate   ART19
What is "critical race theory" and should it be banned?

THE DEBATE PODCAST
Newsweek

OPINION

President Biden Must Do Much More to Make the U.S. a Safe Refuge
BY PAUL O'BRIEN AND ELEANOR ACER

The Positive Impacts of Biden's Capital Gains Tax
BY VANCE ROUSH

Yes, Inflation Is Really Back
BY CONNEL FULLENKAMP

The Real Radicals Aren't in Washington
BY RYAN GIRDUSKY

China's Approaching Demographic Crisis
BY JIANLI YANG

The Big Tech Intifada
BY RABBI YAAKOV MENKEN

How to Win Back Americans Turning Away from Doctors for Medical Guidance
BY KAREN STRAUSS

Joe Biden's Systemic Socialism
BY RICK SCOTT

We on the Left Need to Fix our Broken Israel/Palestine Politics
BY ARASH AZIZI

Four Months After an Insurrection, It's Time to Move Forward—Not Move On
BY CHRISTINE TODD WHITMAN, NORM EISEN AND JOANNA LYDGATE

A SPECIAL MONTHLONG SERIES:

Gen Z: Inspired, Activated & Engaged
April 2021

ASP | Newsweek

SPONSORED INSIGHT

Live and Work in Bermuda

The beautiful island nation is positioning itself as the perfect base for top professionals seeking to embrace the emerging trend of home working.

GET THE BEST OF NEWSWEEK VIA EMAIL

Email address

FREE SIGN UP ›

New $25.41 | Staples multiuse 8-ream case   30% back in Rewards on all Ink & Toner   Business Purchasing Programs

Staples   Products   Deals   Services   Search

STAPLES REWARDS

STAPLES REWARDS
Learn about Free benefits & savings.

FREE DELIVERY, NO MINIMUM
Enjoy on online orders.

UP TO 5% BACK IN REWARDS
In store and online.

MEMBER EXCLUSIVE DEALS
Additional savings, coupons and more.

UP TO
$100off
select chairs and furniture.

Staples Rewards®
FREE DELIVERY, NO MINIMUM
Plus member exclusive deals and more. Join for free

LIMITED TIME

Only $25.49
for Staples® multiuse paper, 8-ream case.

Only $36.99
for Hammermill® Copy Plus® paper, 10-ream case.

20% back in rewards
on all ink and toner.

As low as $114.99
for select printers.

Only 79¢
for hand sanitizer, 8 fl. oz.

Trending now

Rewards
Join today and be rewarded online and in store.
Join today

Orders
Look here for updates on your recently placed orders.

Lists
A powerful tool to keep you organized and make re-ordering easy.
Create a new list

The navigation should **stand apart** from the page visually and appear in some way to be clickable.

Navigation

Often included in the header of a web page is the navigation, or menu, of pages available on a site. The navigation should stand apart from the page visually and appear in some way to be clickable (or tappable in the case of mobile). As discussed in the previous chapter, navigation is an essential part of the usability of a site, therefore the button labeling should be clear and legible.

Often, there is a need to break up the navigation into primary and secondary navigation areas. The primary navigation should lead to the pages most useful to the users and the labels should clearly and concisely convey the content they lead to. The secondary navigation usually contains things like company info, contact information, and possibly a link to a blog or other secondary items. This division not only helps with the usability of a site but it also helps create a sense of visual organization and hierarchy on a page.

The heavily stylized drop-down menu on this site helps it fit in with the rest of the site design.

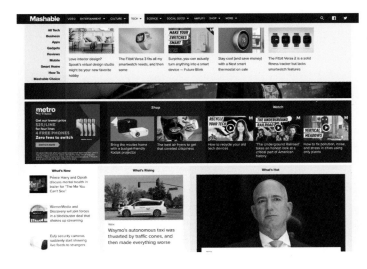

Navigation can be horizontal along the top of a page or vertical along the side of a page, or even a combination of the two. Navigational elements are also contained in the footer of a page, which is discussed later in this chapter.

Mashable.com has an elegant drop-down system that presents the user with photos for the key stories, attracting them to the content.

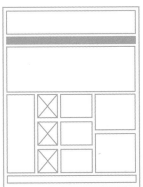

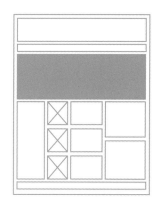

Feature Area

One indication of effective design is a clearly defined hierarchy of information. To achieve this, designers use a focal point—an area in the composition that is perceived before all others and serves as an entry point into the layout. In web design this is often the main feature area. This area usually takes up a large portion of the home page, has the most vibrant color and typography, and usually features some sort of motion or animation. All of these things combine to make it the most important visual item on the page.

The most common option for a feature area is a slideshow of imagery and content from the site. This can be achieved using SEO-friendly technology like JavaScript and Ajax.

The Pixar.com feature area dominates the page and fills the user's screen. The simple design of this page, consisting primarily of a feature area and little else, fits with the clean look of the Pixar brand and focuses the user's attention on a limited number of items.

To achieve hierarchy, designers use a **focal point**—an area in the composition that is perceived before all others.

Apple.com uses the feature area to highlight their latest products. Dramatic photos combined with simple, pithy headlines set in minimalist typography result in an impactful presentation with a clear focal point and call to action.

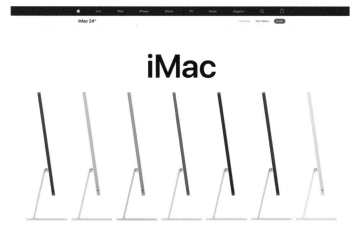

The feature area on Disney.com extends beyond the confines of the box in which it is contained. The background color changes based on the content of the "slide," giving the entire page a unique feel with every change.

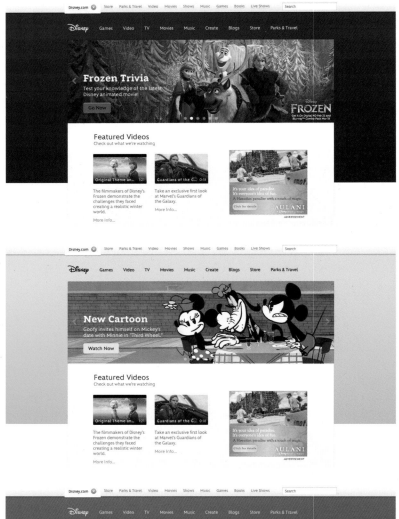

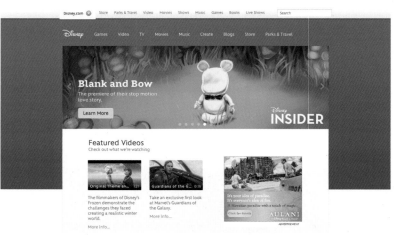

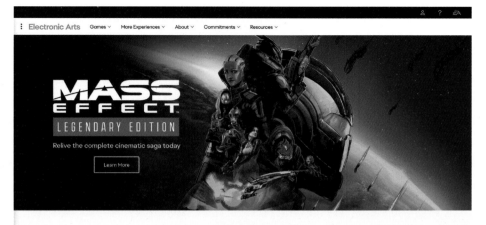

Crisp and precise, this clean design of EA.com employs a very simple and conventional layout.

Featured Games

Latest Games

Latest Updates

EA News   EA Play   Madden NFL   Apex Legends   FIFA   Star Wars   The Sims 4   UFC   Inside EA

Mass Effect™ Legendary Edition
May 13, 2021

Thank You

Launch, #MyShepard, and Thank You

Electronic Arts Inc.   **May 11, 2021**

Electronic Arts Reports Q4 and FY21 Financial Results

Electronic Arts Inc. (NASDAQ: EA) today announced preliminary financial results for its fiscal fourth quarter and full year ended March 31, 2021.

Electronic Arts Inc.   **May 5, 2021**

Welcoming Metalhead Software to the EA Family

Today, we're incredibly excited to welcome Metalhead Software, the talented makers of the fan-favorite Super Mega Baseball games.

Breaks in the content allow users to scan the layout quickly and give them **multiple entry points** into the page.

Body/Content

The body or content area of a website is where users spend most of their time, as it usually represents the end of their search for content. This is where traditional design ideas of legibility and clarity come into play, but with some added considerations. A web page can be any height—however, it is always seen through a window, the size of which is determined by the user's settings. The area of a page that users first see in their browser window when a page loads is the area known to be above the fold. The content in this area must quickly convey the nature of the content that appears outside of that area, known as below the fold. Telegraphing the content on a page with clear and descriptive headlines as well as appropriate imagery is not the same as delivering all of the content above the fold. Users have become accustomed to scrolling down a page in order to reveal more content, just as newspaper readers will leaf through a paper as a story progresses.

It's important to break up long stretches of content with white space and clearly identifiable subheadings. These breaks in the content allow users to skim the page quickly, and it gives them multiple entry points into the content. Dividing up the content by using heading tags (<H1>, <H2>, and so on) helps search engines evaluate the content of a page. Some search engines place a higher value on words contained within these tags, since they tend to summarize the key points from the content. Learn more about SEO in chapter 7.

Colored subheads, iconography, and generous white space make this page from Apple.com easy to scan to find the information you're looking for.

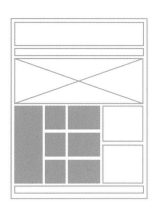

Linked words within the text of a page help to organize ideas and reduce the need for long pages; if a user would like to know more about a related topic, the user can click to another page rather than have all the information on a single page.

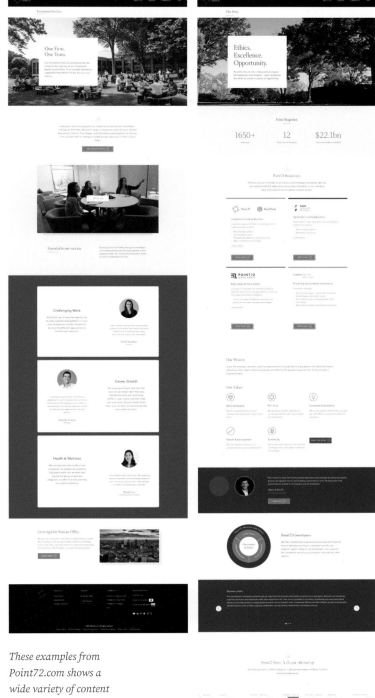

These examples from Point72.com shows a wide variety of content types that serve to break up the page and make consuming the content much easier.

The **optimal line length** for ideal legibility is no more than two to two and a half alphabets—fifty-two to sixty-five characters.

In addition to not having a height limit, web pages also don't have a limit to how wide they can be. Web designers have two options for addressing the problem of page width variability. Most current sites have a fixed width frame or boundary, whereas the content is confined to a box with a set size that floats in the browser window as it expands and contracts. The second option is to have variable-width columns. Variable-width layouts were popular in early web design primarily because they were easy to produce. Designers would simply flow copy into a layout, unconcerned with the consequences of expanding browser windows. The issue with variable-width layouts is that without limits to the length of a line of text, it can become illegible. Typographically, the optimal line length for ideal legibility is no more than two to two and a half alphabets—fifty-two to sixty-five characters. This prevents eye fatigue, both from lines that are too long, where users might lose their place, or lines that are too short, where the user is continually going to the next line after just a word or two.

Jaded zombies
acted quietly but
kept driving their
oxen forward.

These three examples of text show how a short line length (top) and a long line length (bottom) make text difficult to consume quickly. The middle example contains fifty-two to sixty-five characters in a single line, presenting optimal legibility.

Jaded zombies acted quietly but kept driving their oxen forward. The wizard quickly jinxed the gnomes before they vaporized. All questions asked by five watched experts amaze the judge. Six boys guzzled cheap raw plum vodka quite joyfully.

Jaded zombies acted quietly but kept driving their oxen forward. The wizard quickly jinxed the gnomes before they vaporized. All questions asked by five watched experts amaze the judge. Six boys guzzled cheap raw plum vodka quite joyfully. Just keep examining every low bid quoted for zinc etchings. Sixty zippers were quickly picked from the woven jute bag. Few black taxis drive up major roads on quiet hazy nights. Six big devils from Japan quickly forgot how to waltz. Painful zombies quickly watch a jinxed graveyard.

Wikipedia.org uses a variable width for the body/content area of the page. Both of these images are of the same page, showing a narrow browser window and a very wide window.

Sidebar

The sidebar of a web page contains secondary information that either supports the main content of the page or directs users to related content through the use of sub-menus and links. Areas of a sidebar are often sold for advertising space. Skyscrapers and Big Box ads, as they are known in the online media industry, typically fit well within the modular structure of a sidebar. As with the header, the design of a sidebar should blend in with the look of the site so as not to visually overshadow the content of the page, helping to create an overall feel for the page.

Sidebars, like the one shown here from NewYorkTimes.com, are useful for providing supporting information, related content, as well as advertising space.

The New York Times

MUSIC Allison Russell Faces Her Past in Song

Allison Russell Faces Her Past in Song

The singer and songwriter's debut solo album, "Outside Child," tells a harrowing story with a survivor's joy.

Allison Russell sings plainly about the sexual abuse she endured as a child on her debut LP. But it's an album of strength and affirmation, not victimization. *Bethany Mollenkof for The New York Times*

 By Jon Pareles

May 13, 2021

It took a long time before Allison Russell was ready to sing her own full story. Once she was, the songs came rushing out.

Her solo debut, "Outside Child," speaks bluntly about sexual abuse by her adoptive father. She spells it out, over a steadfast Memphis soul beat, in the first song she wrote for the album, "4th Day Prayer": "Father used me like a wife/Mother turned the blindest eye/Stole my body, spirit, pride/He did, he did each night."

Yet in that song and throughout the album, she also sings about deliverance and redemption, about the places and people and realizations that helped her survive and claim her freedom. It's an album of strength and affirmation, not victimization.

"When you're around her and her family, she just is pure joy," said the singer and songwriter Brandi Carlile, who got to know Russell after hearing and admiring the album, due May 21. "Her smile stretches from side to side of her face, all the time. And you would never know that she came from a brutal and harrowing childhood situation, except for the fact that she honors it by telling you."

Editors' Picks

A.O.C. Had a Catchy Logo. Now Progressives Everywhere Are Copying It.

Paulina Porizkova, Full-Frontal Emotion

To Understand Amazon, We Must Understand Jeff Bezos

RED POST: VISIT FLORIDA
There's a Fun, Floating Family Vacation in Your Future

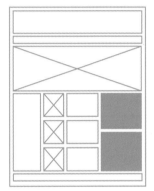

A sidebar can also be used for navigation or filtering. And as seen in this example from katevassgalerie.com it doesn't have to be on the right side. Any sub-content that isn't part of the body of the site is considered a sidebar.

Footer

The footer, or the very bottom of a web page, is a critical part of web design, performing tasks for both the user and search engine optimization. In the early days of web design, the footer would contain the copyright information for the site as well as a couple of links. Over time, web page footers have grown to resemble a mini–site map, with links to each of the main pages of the site. These links not only help the user navigate the site but also help search engines like Google index the site properly, improving the search engine ranking—Google places a higher value on words contained within links.

Technically, the footer of a website contains much of the specialized coding for the page like page-tracking code or lengthy JavaScript functions. This is again due to SEO. Long bits of copy at the top of a page will push the important information down farther. Google places a higher value on information that's higher up on a page.

The footer of a website contains many items, including navigation, contact information, featured content, and a call to action to sign up for a newsletter.

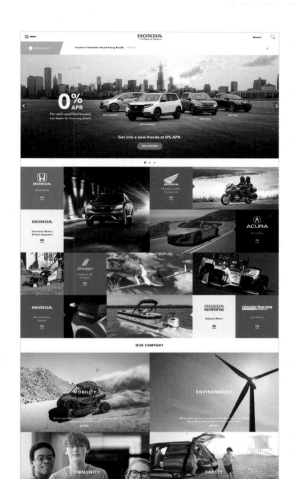

The footer of automaker Honda.com gives a complete list of areas of the site and also adds a small bit of visual interest with a randomly-appearing image.

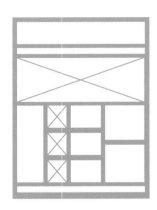

Background

In the earliest days of web design, designers would use a repeating graphic in the background of a web page, imitating the effect of patterned wallpaper. Today, thanks to increased bandwidth and faster connection speeds, web page backgrounds are used in bolder and more sophisticated ways to complement the content of the page. Backgrounds can be used to create depth or dimension, add richness with texture and color, or even expand the content beyond the borders of the page.

The designers of cpm-interiordesign.be turned the background into a critical element of the page. Large textural images lie beneath a simple CSS design structure.

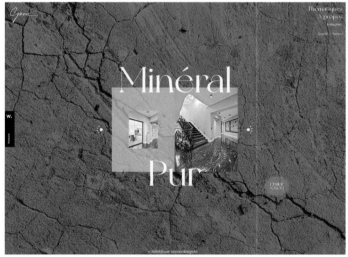

Funding and supporting Europe's finest game companies

Nordisk Games provides smart growth capital, strategic guidance and operational support

Who is Nordisk Games?

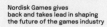

Nordisk Games gives back and takes lead in shaping the future of the games industry

Bornholm Game Days

Supporting the ecosystem

Diversity in games

The Egmont Foundation

Climate Agenda

The large background images on Nordisk Games (right) integrate with the content and give added detail to the content.

outofthevalley.co.uk (left) uses alternating colored background (which were difficult to capture) to help guide the user down the page, giving each section unique visual interest.

The background images on en.opera.se give a dramatic sense to the pages of the site because of their contrast of scale.

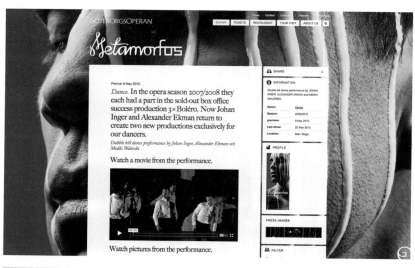

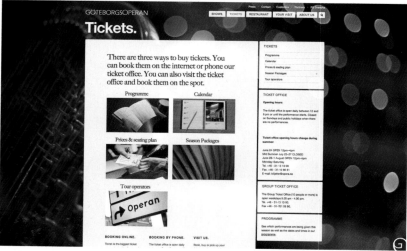

There is some debate between web designers and usability experts regarding the use of dark backgrounds and light text. Most experts believe that it's more difficult to read light text that's reversed out of a dark background; however, many designers prefer the look of dark backgrounds. As with many other decisions a web designer makes, this one comes down to the tolerance and preference of the user.

The dark photograph in the background of this site gives the pages texture and creates a mood that's dark, rugged, natural, and masculine.

CHAPTER 5

The Elements of Web Design

Creating a design system so that dissimilar types of content appear to work together is what graphic designers have been doing for centuries, and web design is no different. The previous chapter explored various structural and spatial methods of organizing space and creating a structure. Design is about more than simply organizing information, however; it's about making something distinctive and memorable. This chapter explores the aesthetic treatment of the elements within a design that not only help form relationships within a system but create a visual style.

Web Design Style

This page from the Gutenberg Bible, the first Western example of movable-type printing, represented state-of-the-art technology when it was produced in the 1450s.

A design style is an attempt at connecting with a user's sensibilities and a basic need to relate to things. The elements of a design style include color, texture, typography, and imagery use. Additionally, there are means of manipulating these elements, including creating a sense of scale or depth, animation, and variation. The crafting and manipulation of these aesthetic elements of style make a particular design unique and, better yet, memorable.

In all forms of design, a style comes primarily from two areas: the trends of the time—what's fashionable—and the technology that's available to create a piece of design. Graphic design, which dates back to cave paintings and carries on through the carved letterforms on Trajan's column, handwritten manuscripts, Gutenberg's movable type, right on through to photo reproduction and the modern computer age, has always been heavily influenced by the technology available to produce it.

The same is true in web design. As computer technology, browsing software, and the markup language that makes up a design become more advanced, they influence the design styles and trends. Through it all, however, great design is defined by the fundamental understanding of the hierarchical structure that makes up a layout, explored in the previous chapter, combined with the elements of style that give a design its uniqueness.

Color

More than any other design element, color has the ability to guide, direct, and persuade a user. In addition to its instructive qualities, color can appeal to a user's emotions by setting a mood or a tone for a piece of design. Colors signify meaning for many people and cultures, making it a powerful tool for designers. The immediacy with which color can be recognized makes it valuable for forming clear relationships.

Color has three main properties: hue, which is commonly known as the color; value, which is the darkness or lightness; and saturation, which is the vibrancy of a color. Because web design is based on the colors of light (red, blue, and green), the range of colors is greater than with print design, which uses the reflective palette (cyan, magenta, yellow, and black). Although there's a broader color palette, predicting the exact color a user sees is difficult because of variations among monitors and operating systems.

Color is used in BobbyFlay.com as a device to emotionally connect with a user. The bright, vibrant colors are intended to excite and engage the user.

Sometimes limiting the number of colors used in a layout can have a big impact. The use of yellow in the navigation, design elements and photography helps Meacuppa.be create a strong and memorable visual identity.

Relationships of color help users create associations among otherwise unrelated elements within a design.

Color is an excellent way to create relationships within a design. USAToday.com uses colored type to signify various categories; blue for news, red for sports, purple for life, green for money, etc. This use of color helps users quickly scan a page to find information without the need for a lot of reading.

Contrasting color can help a designer guide and direct a user through a layout.

Effectively using color doesn't necessarily mean creating a colorful design. This example, NewYorker.com, uses only touches of red among a sea of black and white to lead the user and highlight key information. The schematic (above right) illustrates how color guides the eye down and around the page.

Texture

Adding texture to a web design gives the user the sense of a tactile experience and helps connect him or her to the content of a page. Types of texture can range from smooth, shiny buttons that are common in web 2.0 design, to rough or grungy treatments, to type imagery or backgrounds. Aside from the stylistic treatments of texture, it's important to remember that on a macro level, every design has a texture, intended or not. Type, images, and illustrations combine to make an overall texture that the user perceives on a subconscious level.

The stucco texture on Jarritos.com comes from the Mexican theme that permeates the site.

Adding texture to a web design gives the user the sense of a **tactile experience** and helps connect him or her to the content of a page.

The fabric texture that is subtly visible in this sample gives the page added interest and depth.

Handiemail.com uses a subtle paper texture to enhance the concept of hand-written letters made from emails.

Imagery & Iconography

Studies show that users don't read websites, they scan them. For web designers, the use of images or iconography can mean replacing wordy descriptions with single images, making a layout easier for a user to get information from quickly. A designer's choice of imagery should be deliberate and add to either the branding or the message of the page. All images add to a web page's weight or file size, so gratuitous use of generic imagery can impede a good user experience.

This scrolling page is illuminated with fun and engaging illustrations that blend content and decoration nicely.

Fixate.it tells an entire brand and product story with imagery. The images convey the features of the product, likely better than paragraphs of copy could.

The icons seen here on ThemeKingdom.com help the user quickly scan the content by relating visual elements to the text.

Icons are used in this example to quickly convey the content of the article so users can scan the page and glean a lot of information.

mailchimp

Log In   Sign Up Free

Mailchimp 101

New to Mailchimp and not sure where to start? We'll walk you through
the basics so you know what to expect along the way.

Let's start building your audience

Just by bringing all your contact data into Mailchimp, we can
start to show you helpful insights about your audience. We'll
help you find new ways to talk to people who love your brand
—and new ways to reach people who are likely to.

Even if you don't have contacts to market to yet, you can
always log in and start playing around with our design tools. In
fact, we recommend it.

Get started

Want to learn more? Here are a few resources to help.

Getting Started with Your Audience   →

Import Contacts to Mailchimp   →

Requirements and Best Practices for Audiences   →

*Hand-drawn illustrations that appear on
Mailchimp.com bring unexpected visual
delight to each page.*

Both of these samples, one from Apple.com the other from Harrys.com, use large photographic elements juxtaposed against small text to create a sense of scale. In the case of Apple, this highlights a key product feature — size. With the Harry's example, scale creates a heroic feel for the product.

Scale

Contrast of size or scale is one way designers can add a sense of drama to a design. Having a dominant element is critical to creating a clear sequence or hierarchy of elements within a design. Scale is a relative design element, so in order to achieve a dynamic feeling of scale, small elements must be included in the layout for comparison's sake. Large design elements that break out of borders or even bleed off the page also heighten the sense of scale.

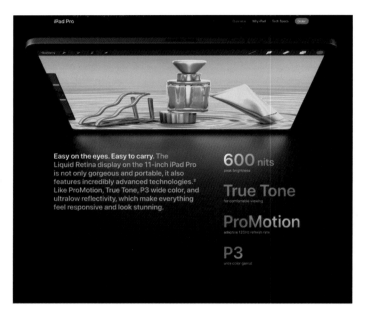

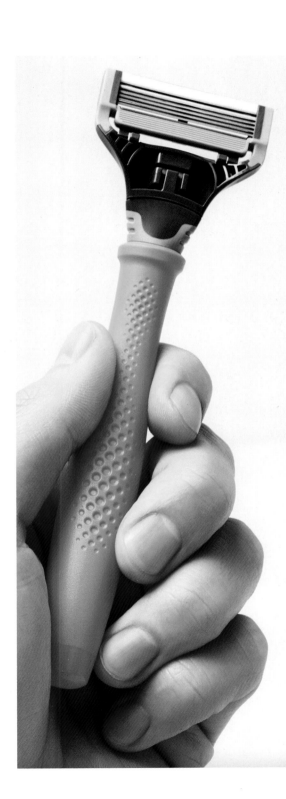

Handsome Razor, Increased Control

The Truman handle is designed with a rubberized matte exterior, texturized grip pattern, and weighted core for maximum grip and control. It may be the best-looking thing in your bathroom (other than you.)

Depth & Dimension

Applying depth and dimension to a page gives it an element of realism, and, like texture, gives the user a more tactile experience. There are many ways to create the illusion of depth in a web design, like simple overlapping of design elements, adding gradient color and shadows, or even creating three-dimensional elements. Adding depth to a web page can help add visual interest and draw a user into a design.

This web page for developer Oliver James Gosling gives new meaning to the phrase "above the fold." The subtle gradations of gray and cast shadowing give the appearance of an unfolded brochure.

The sites pictured here use overlapping design elements and shadowing to create the illusion of depth. These dynamic layouts engage the user and encourage exploration.

Adding depth to a web page can help add **visual interest** and draw a user into a design.

From three-dimensional type and objects in perspective, to layered elements and subtle gradations of color and shadowing, Syfy.com appears to be completely designed around the concept of depth and dimension. Almost every element of the design seems to lift off the screen. The main feature area is a shelf where elements stand, casting a shadow onto the other pieces of information.

Animation

Animation is a tool used by digital designers to layer information, create a sequence of information, or simply surprise and delight the user. Animation can be the focal point of a design— like a slideshow or video in the main feature area—but animation can also be simple and subtle, like small amounts of movement when a user mouses over a button. Too much repetitive animation, especially on pages with a lot of content, can become distracting to a user. Web design best practices dictate that the designer should always give the user the ability to pause a large animation, or, if an animation is looping, to cycle for no more than three cycles.

User-initiated animation is a great way to provide feedback or build a story. Animations can be triggered by a user clicking a button or by scrolling down a page. These touches, if used in ways that are true to the client's brand, can really heighten the feeling of an interactive experience rather than a passive one.

The designers of Cyclemon.com create a sense of motion with a scrolling effect on their website. The is contextually relevant for a bicycle company.

Another popular way to add realism and depth to a page is with a parallax scrolling effect. Parallax is a physical phenomenon in which objects that are far way appear to move more slowly than objects that are closer. The parallax technique has been used by makers of cartoons and video games for years. Web designers now use a parallax scrolling effect to add interest and depth to a page.

A stunning and delightful example of scrolling animation is found at WeeSociety.com, designed by the design firm The Office. As the user scrolls down the page, elements animate and build to create colorful and playful vignettes.

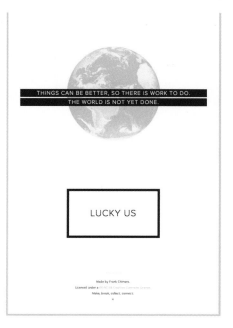

This stunning site by Frank Chimero is an outstanding example of the various effects that can be created by parallax scrolling, from multi-part animation at the top, to the illusion of motion pictures in the middle, to more subtle effects at the bottom.

Modularity

Modularity can mean a couple of things when it comes to web design. For a web designer, modularity means creating reusable or modular design assets that fit within the established grid system and get reused throughout a site. These modules not only create design efficiencies, but they also help with usability by repeating recognizable elements that a user can remember.

Modularity can also refer to the necessary design flexibility required in web design. Some types of websites, like news portals, need to accommodate varying lengths and types of content from day to day—even from hour to hour. Therefore, web design systems for these sites must be flexible to expand and contract as the needs of a site change. This isn't a web-specific principle; newspapers, magazines, and even corporate identity systems need to have an element of modularity to be effective. What is unique about the web is the speed with which items in a design need to change (which makes planning ahead an essential part of web design), the fact that the user can sometimes control the content, and the need for expansion and contraction, making the ultimate outcome unpredictable. Sites that have user-controlled modularity use JavaScript technology to enable users to drag and drop content "blocks" above and below the fold to create their own hierarchy of information.

The modular elements of this site work perfectly from section to section, page to page. This interchangeable design system makes management of the content much easier for the site administrators.

< Resources

Design Education Resources

AIGA supports education throughout the arc of a designer's career, including special programs for educators, a group critical to advancing the profession.

AIGA Design Teaching Resource

The AIGA Design Teaching Resource is a peer-reviewed platform for educators to share strategies for teaching historians, technicians, and process reflections.

Browse Teaching Resources

AIGA Design Educators Community

The AIGA Design Educators Community is comprised of Collaborative educators providing a voice and leadership to the Communities

Meet the committee →

Design Futures

This research project examines some trends shaping the context for the practice of design. This change is the nature of work calls for new skills and perspectives to become traditional college level education. It is critical that the industry expands its knowledge and expertise in certain to continually native and professionally motivate or in progress for changing client demands and new opportunities for design influence.

Read the trends →

Dialectic

Dialectic seeks to publish scholarship, and must study such a vision that will enlighten and inform a diverse audience of design educators engaged not only in classroom teaching experience but also in differing forms of research and professional practice. It is open access studies find strict movements and the official record of the AIGA Design Educators Community (DEC).

Access past issues →

Dialogue

Dialogue is the ongoing action of code-sharing the content to the conferences and annual programs of the AIGA Design Educators Community (DEC) mixer of Dialogue scholarly papers from DEC conferences that focus on topics affecting design education, research, and professional practice, although each conference varies in feature. Michigan Publishing, the hub of scholarly publishing at the University of Michigan, performs Dialogue on behalf of the AIGA DEC.

Access past issues →

NASAD and Accreditation

The AIGA NASAD briefing papers address various aspects of graphic design education and are intended to serve a number of different audiences

Read the briefing papers →

Promotion and Tenure

The purpose of this resource is to provide guidance to those involved with Promotion and Tenure (P&T) processes of Graphic Design and Visual Communication Design Educators at all Universities at higher education. It is not meant to address all possible topics and issues related to the P&T process, but should assist in dealing with issues commonly encountered in the P&T process, and will provide suggestions on which policies and procedures may be based, at the discretion of the institution.

Read more →

Professional Standards of Teaching

A design educator influences values that demonstrates respect for students, other educators, academic institutions, the profession, the public, society and the environment. These standards define the requirements of a design educator are expected in the classroom of an AIGA member teaching range.

Read more →

Guide to Internships

AIGA believes that quality internships provide an invaluable stepping stone towards professional practice and careers intimately within the design profession. We found those who open their doors to young designers and generously share their knowledge and experience with the next generation of design practitioners.

Read more →

Bibliographies

These bibliographies for design educators and students address six eclectic areas of topics: Cognition and Emotion, Cultures, Studies, Design Planning, Education and Learning, Theory and Interaction, and New Media Design.

View the bibliographies →

High School Design Curriculum

This four unit graphic curriculum has been created specifically for high school educators. From addressed include An Introduction to Graphic Design, 2D design Basics, Design Process, and Typography, as well as a general and supporting functions. Created by AIGA. This course will navigate from AIGA Directory.

Download the curriculum →

SHIFT 2021

The turn-migration work series to gather the design educator community to take most of email we are, what we have learned, and what we want to do next. The summit will focus on themes of Teaching, Research, and Community. More information to come for 2021

Learn More

Additional Teaching Resources

Course Planning   +

In the Classroom   +

Student Resources   +

Select Publications   +

Related Videos

Student-Run Design Studios / Concepts of Affordance / International Design Collaborations / The Value of Design Education

---

Design

AIGA, the professional association for design, advances design as a professional craft, strategic advantage, and vital cultural force.

Explore

Leadership	Journey & Career	Design Practice	Business
Design leaders look to standards and thinking through thought leadership, employment, and techniques. The future of design.	The design journey is not linear. Series discusses your career paths, salary techniques, and the emerging profession.	Discover our collections and learn about different perspectives of design practice.	Design is a critical strategic advantage to business through independence, exposure to diverse voices, experiences, and perspectives.
Strategic Design & Innovation	**Education**	**Research & Insights**	**Conferences**
Strategic design helps organizations to use all of the resources to innovate problems solving, developing opportunities, applying methods, and developing solutions	Design education is one of the cornerstones for aspiring designers and professionals who pursue their careers.	Here's where that where is done where where to navigate, find studies and results, that help our thinking and outcomes.	Joining the best-known conferences and research annual. Explore conferences that will later you a deep sense of design.

What's New

AIGA Design & Business Conference

Join us for the 2021 AIGA Design + Business Conference May 12–14, where you will take a deep dive into the creative process and gain actionable insights from impactful campaign along with the stories that launched them.

Learn More

AIGA Design Point of View Research

The Design Point of View (POV) Research Initiative provides a deep understanding of design and the people who engage with it to reveal the underlying challenges of the profession, guide future impact, and influence policy and decision making. Download the Executive Summary and learn how you can participate in the Design POV.

Learn More

Keep in Touch with AIGA

Signing our email content preview is the the first for updates on exhibitions, events, programs, and stories and more

Sign Up

Find a Chapter

Searching to access something through your local chapter and a vast variety of local and national programs and events

Learn more

"Being a member of AIGA has given me access to educational resources for my professional development and business opportunities...it's opened the door to meaningful relationships with a diverse group of dedicated people with whom I'm proud to be in community with.

John Hornsby
AIGA Advocate

Design Content at Your Fingertips

Shop AIGA for design inspiration and get the free digital download, "What Their Color" book You In Design School to have access to AIGA for an design on your bookshelf

Get the Magazine

Next Up: Resources

Business & Freelance Resources	Design Education Resources	Student Resources
A labor one selection of resources, legal guides and business planning resources to increase a design industry competencies	Created by our Design Educators Community (DEC), explore the collection of curated resources for design educators.	Find the help you need to navigate the design industry post-graduation.
Explore →	Explore →	Explore →

Engage & Learn	Partner	Connect
AIGA Membership	Career, Jobs & Projects	Sign up for our newsletter
AIGA Design + Business Conference	National Partner Program	Subscribe
AIGA Design Conference	Sponsorship & Advertising	
AIGA Design Educators Community	Support Us	AIGA address
AIGA Design POV Research	Contact Us	New York, NY 10000
AIGA Portfolio Festival		
AIGA Webinars		
More		

© 2021 AIGA . Terms & Conditions . Data User Testing

Variability

The speed at which a web designer can apply changes, combined with the need to continually refresh the look of a site, gives web designers the ability to vary elements of a design based on things like sections of the site or specific events—or randomly. What was once considered unthinkable—altering a corporate logo, for example—can now be a playful way to add relevance to a website. The best way to keep a site fresh is by updating the content. But if that's not possible, design variations can give the user the impression that a site is fresh and current.

The USAToday brand is based on the variable utilization of the circle. Once a globe, it now represents all of the various topics reported by the news company.

What was once considered to be **unthinkable** can now be a playful way to add relevance and variety.

166

These images from Google.com show the
playfulness with which their designers treat the
Google branding, from the anniversary of the
moon landing to Dr. Seuss's birthday.

CHAPTER 6

Web Typography

" Typography is the one area in graphic design where there are truly rights and wrongs; there are better-thans and there are randoms.**"**

Alexander W. White
Chairman Emeritus, The Type Directors Club

Why Type Matters

Typography, of all elements of design, can have the greatest effect on the success or failure of a piece of communication. This is because type carries the message, and the craftsmanship of the typography can either enhance or take away from the message. Many designers share a passion for the art of typography and can spend hours kerning letters, adjusting the rag on a column of type, or hanging punctuation. With web design, however, this level of finite control is difficult, or in some cases not possible at all. But before examining the specific nuances of web type, it's important to understand a few universal principals of typography.

In historical terms, a font is a complete set of characters that make up a single size, style, and weight of a typeface. The term *typeface* refers to the unique styling applied to a set of glyphs, including an alphabet of letters and ligatures, numerals, and punctuation marks. Due largely to their use in relation to computers, the two terms have evolved to be interchangeable. The term *font* no longer refers to a single size or style, and can even refer to the digital file used by the computer to display typefaces.

It has been said that great typography is invisible, but that's only half the story—typography can also be beautifully expressive and attention-grabbing. In either case, type must carry a message to the user. The two opposing characteristics, which combine to attract a user and convey a message, are called readability and legibility. Both are essential for effective communication.

Readability refers to how well type can attract a reader. Typographic posters, book covers, packaging, logos, and magazine features, for example, must have a readable quality to them in order to get the attention of a reader—a quality that makes a person stop and want to read. Readability can come from size, font usage, composition, color usage, abstraction, or anything that helps type—or, more specifically, the message—stand apart. Effectively readable type expresses meaning through form beyond the content of the words it displays. The FedEx logo is an example of this idea. The bold, geometric shapes of the letterforms imply stability or reliability, while the negative-space arrow between the capital *E* and lowercase *x* implies forward movement and speed—all this with the use of only five letters and two colors.

Legibility, on the other hand, references the ease with which a reader can gather a message, especially when it comes to long stretches of copy. The recognizability of individual characters in a font as well as type size, leading, letter spacing, line length—even color and backgrounds—play a role in how effectively legible type appears. Truly legible type makes it possible for the reader to perceive only content and not be distracted by formatting or decoration.

These two aspects of type play a big part in effective web typography; however, the level of control a designer has and the methods he or she uses to achieve them can be very different. Readable or expressive typography can be important on the home page to grab the user's attention, define a unique brand characteristic, or alert the reader to a site feature or special offer. Legible type is essential for article or blog text and can make the difference in the success of a site that invites users to return to read long articles or posts.

House Industries offers hundred of unique fonts for every application. The selection of a typeface can have a significant effect on the overall feel of a web project.

Measuring Type

Type and typographic properties such as spacing are commonly measured in em units. An em is a square unit that represents the distance between baselines when type is set without line spacing or leading. An em square is equal to the size of the type; for example, an em space for 12 pixel/point type is a 12-pixel square.

While an em is equal to the type size, the individual characters don't necessarily fit within an em square—they can be larger or smaller. As seen in the diagram below, a Dispatch M fits within a single em unit of 110 points; however, the Burgues Script M at the same size is not confined to the em unit.

In web design—more specifically, CSS styling—type can be defined using em units. Ems are used for relative sizing and for type they're used in the font-size attribute.

Most browsers default to 16-pixel type as a general rule. So if a designer specs type at 75 percent, the size of the type will be 12 pixels. The default can also be altered globally by styling the <body> tag. If in the body tag the font size is set to 62.5 percent, then the default for all type on a site is 10 pixels (16 x .625 = 10). Therefore, the math for defining other sizes becomes easier: for 15-pixel type the font size would be set to 1.5 em; 24-pixel type would be 2.4 em, etc.

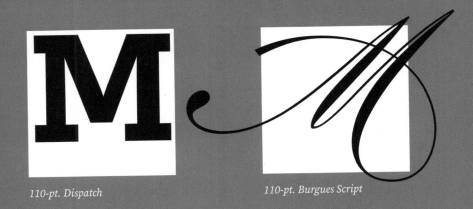

110-pt. Dispatch

110-pt. Burgues Script

The optimal choice for displaying type depends mostly on the **needs of the client** and the **capabilities of the target user.**

While there are limitations to the control a designer has over typographic details on the web, there are also methods, unique to web design, of turning over control to the user so he or she may create personalized settings for legibility. Many sites give users the ability to change the size of text, and some sites even give users the option of choosing their own fonts.

Designer Control

User Control

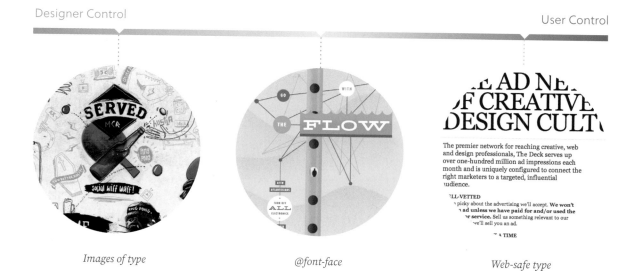

Images of type

@font-face

Web-safe type

Types of Web Type

For years web designers were constricted to only two options for web typography—images of type and system fonts (the fonts found universally on devices used to browse the web). Using images to display type is a static method of rendering type—the type is rendered once by the designer or producer, and that image is distributed throughout the web to be viewed by the user. Using web-safe fonts is a means of displaying content as live text, which is rendered by the user's browser. Live text generally offers less control to the designer but more control to the user with which to manipulate aesthetics and/or search the content.

However, that all changed with three important innovations: widespread browser support for the @font-face CSS command; the emergence of font delivery technology; and the development of the Web Open Font Format (WOFF). Now designers have three primary choices when selecting methods of displaying type within a design:

- Images of type
- Web-safe system fonts
- @font-face fonts

Why is this choice so significant? The reason selecting a method with which to display type is so critical to web design is due to the fact that type delivers content and content drives the success of most websites. Content is what users search for. Content is what search engines index and catalog, and search engines can only pull content from live text—images of type are not indexable by search engines. Content, however, must be dressed with some form of style or branding in order to be truly effective for the website's owner. Purely displaying content without some sort of visual expressiveness or uniqueness decreases its memorability and therefore decreases its value to the client. The following pages explore examples of each method of displaying type, and details the benefits and drawbacks for each.

Image Type

Images of type offer a web designer the most control over the typography on a web page. A designer can freely choose a font from his or her library, adjust the kerning, add filters and effects, etc.—all the things that traditional print designers are used to doing with type. Images of type enable a designer to match branding requirements for a client exactly, or to express a concept precisely as the designer (or client) envisions.

There are a couple of significant drawbacks with this method of displaying type, however. All-image websites, where the type is rendered as a jpg, png, or gif image, are extremely limited in their ability to be indexed by search engines, and thus limited in their ability to be found by users. While it's possible to include searchable content within the alt tags—a tag within the image tag that allows the webmaster to input text describing an image, used mainly for handicapped accessibility—this text does not have a high value with search engines because it can too easily be manipulated to deceive the user.

This example from AustinBeerWorks.com uses illustrations and images of type to create a layered and rich page design.

Served-MCR.com is rich with hand-drawn type and illustrations as well as animations to create a grungy and whimsical look to this site for a ping-pong tournament for creative folks.

ReadyToInspire.com (below) uses a beautiful combination of hand-drawn letterforms for the headline and the dropcap with web-safe type for the body copy.

The best approach for using images as type on a website is to limit the use to particular areas of display type where the images can have the most visual impact. Commonly, designers choose to use images of type for the main navigation of the site; however, this is particularly damaging, as search engines place a high value on linked content. If the main links are images, the links' value cannot be captured by search engines. The bottom line is that images of type are a great way to add personality or brand recognition to a website but should be used extremely sparingly in order to maintain the searchability of a site.

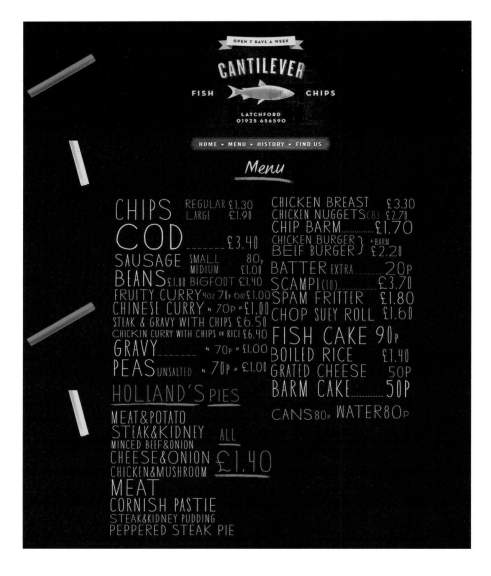

The hand-drawn type in this layout for cantilever-chippy.co.uk conveys the sense of a chalk menu board.

Finger-painted letterforms divide up the sections of this page for VintageHope.co.uk. The distinctive look and feel gives the site a memorability that standard headlines may not have.

The identity for the Branding graduate program at the School of Visual Arts is distinctive, creative, and could not be replicated using only web-safe type. The solution here is to integrate enough of the custom lettering to maintain the recognizability of the brand with web-safe type for legibility and searchability.

Images of type enable a designer to exactly **match branding requirements** for a client, or **express a concept** precisely as the designer (or client) envisions.

Hand-drawn typography defines this site for Chester Zoo. The kid-like feel of the site reflects the audience the designer and the client were trying to appeal to.

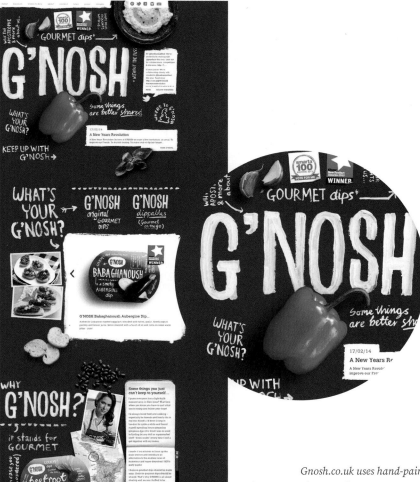

Gnosh.co.uk uses hand-painted type and photography to create a crafted setting in line with the concept of hand-crafted food.

Web-safe Type

With the prevalence of @font-face type available to designers, web-safe fonts seem to be becoming extinct. Perhaps someday, but for now, web-safe fonts still offer two distinct and important advantages:

Web-safe fonts were designed specifically for screen use. Fonts like Georgia or Verdana, both designed by type design master Matthew Carter, were created with the intent that they would be used with back-lit conditions and at small sizes. As a result, they have large x-heights, open counter spaces, and wider letterspacing (see diagram) for maximum legibility.

The @font-face command is a series of code and font files that a browser must load in order to render the type. Therefore, they can slow down the load time of a page. Because of this, it is common to use @font-face type for display type and web-safe fonts for the text or body copy.

Windows	Mac
Arial	Arial, Helvetica
Arial Black	Arial Black, Gadget
Comic Sans MS	Comic Sans MS
Courier New	Courier New, Courier
Georgia	Georgia
Impact	Impact, Charcoal
Lucida Console	Monaco
Lucida Sans Unicode	Lucida Grande
Palatino Linotype	Palatino
Book Antiqua	Georgia
Tahoma	Tahoma
Times New Roman	Times
Trebuchet MS	Trebuchet MS
Verdana	Verdana
Symbol	Symbol
Webdings	Webdings
Wingdings	Zapf Dingbats
MS Sans Serif	Geneva
MS Serif	Georgia

Being able to do **more with less** is an essential skill for a web designer.

This stark and stunning layout for decknetwork.net uses only web-safe type to display the text. The all-caps headline at the top is Georgia, designed by Matthew Carter.

Font Stacks: Designers or coders define web fonts in the CSS with what is known as a font stack. Font stacks are prioritized lists of fonts, defined in the CSS font-family attribute, that the browser will cycle through until it finds a font that is installed on the user's system. Font stacks list fonts in order of the designer's preference: preferred, alternate, common, generic. Common font stacks include:

font-family = Georgia, [if you don't have that then use] "Times New Roman", [if you don't have that then use] Times, [if you don't have that, please just give me something with a . . .] serif;

The limitations and unpredictability of font stacks present a challenge to web designers. Limitations also lead to creative solutions. Doing more with less is an essential skill for a web designer. The sites pictured here represent a wide visual language using only web-safe typography.

Wordpress.org (above) uses a beautiful mix of Georgia for display type and "sans-serif" for body text.

This all-type solution for the Seed Conference announcement showcases many of the possibilities of CSS type styling. Varying type sizes, colors, and alignments create a clear hierarchy within a unified piece of design.

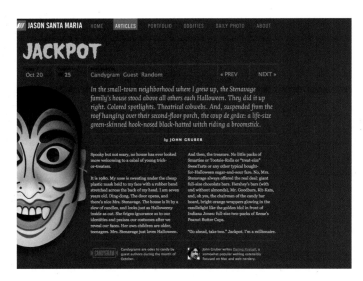

Jason Santa Maria is considered to be one of web design's most creative talents. Pictured here are two pages from his site, JasonSantaMaria.com, where he displays his mastery over type, web type, and imagery integration.

The Anatomy of an Effective Web font

200-POINT GEORGIA

Designed in 1996 by Matthew Carter (hinted for optimal screen viewing by Tom Rickner) specifically for the web.

200-POINT TIMES ROMAN

Designed in 1931 by Stanley Morison and Victor Lardent (Monotype) for the *Times* newspaper.

atf atf

COUNTERS

The larger counters on Georgia increase legibility

X-HEIGHT

Notice the difference in x-height at the same type size

CAP HEIGHT

Even the height of the capital letters differ at the same point size

BASELINE

The line on which letters sit and the starting point when measuring the x-height and cap height

POINTS & PIXELS

The most common unit of measure when dealing with type is points and picas. There are 72 points in .996 inches and standard screen resolution is 72 pixels per inch (PPI). Therefore, one point is equal to one pixel when referencing elements at screen resolution.

The x-height, represented by the blue line, is the distance between the baseline—where the letters sit—and the top of a lowercase letter. It's clear to see that Georgia, designed by Matthew Carter specifically for the web, has a higher x-height than Times Roman at the same size. A counter is the "hole" created in letters like a lowercase a. In Georgia, the counters are larger and more open. These characteristics combine to make Georgia more legible that other serif typefaces when viewed on screen.

ALlistApart.com, uses a beautiful combination of web-safe type (Georgia for the body text) and @font-face (Franklin Gothic for the display type). This cohesive, type-driven layout has defined the A List Apart brand for many years.

tdc.

Shop   Donate   Search   (Join us)

COMPETITIONS   EVENTS   RESOURCES   ABOUT   MEMBERSHIP

RECOGNIZING EXCELLENCE

The TDC celebrates and amplifies the undeniable power of typography. We are a global community united by the shared belief that type drives culture and culture drives type

---

Established in 1946, the TDC today curates a calendar of typographic intrigues designed to:

—build a community through public events and platforms —support the growth of students and early career professionals —recognize excellence in type design across the world.

In connection with our landmark yearly competitions, we produce and publish a Typography Annual and coordinate eight traveling exhibits of award-winning work. We host and sponsor numerous events—conferences, courses, workshops, exhibitions, and frequent Type Salons —both in our New York City home and at supportive institutions near and far.

Graphic designers, art directors, editors, media makers, students, and entrepreneurs, anyone who loves typography, letterforms, and the written word... join us.

Join the TDC

(MEMBERSHIPS)

The Type Directors Club, TDC.org, naturally has some of the most dynamic and striking typography on the web. The site employes a number of typographic elements to create a visual feast for the eyes.

tdc.

COMPETITIONS   EVENTS   RESOURCES   ABOUT   MEMBERSHIP

NEWS

Mag Men Revisited

Anne Quito, Milton Glaser, and Walter Bernard answer your questions on Editorial Design

On May 28, Anne Quito and Walter Bernard gave a presentation for TDC Virtual Salons on their book, *Mag Men*. Milton Glaser was not able to attend but he has responded to some of the attendees' questions. The TDC has compiled their answers to your most salient questions.

OUR HISTORY

1946–1956

1 / 4

The First Ten Years

Founded as the Type Directors Club, the TDC's earliest membership included such luminaries and leaders as **Aaron Burns**, Freeman Craw, and Louis Dorfsman. Our inaugural lecture series, titled Ten Talks on Type, featured Milton Zudeck, O. Alfred Dickman, and Hal Zamboni. It quickly became an annual event.

In 1955, we established the annual competition to recognize outstanding work in the profession, opening it up to non-members in 1956.

InkAndSpindle.com, seen on this page, uses the Google Font Muli designed by Vernon Adams. The minimalist font can be used for both headlines and body text as seen here.

@Font-face Type

186

In truth, the @font-face command existed in CSS2 and dates back to 1998, but there is a problem with it when used by itself. @font-face uses font files located on a server to display a typeface in a browser exactly the same way images appear on a server and are displayed on a page. Therefore, with very little hacking ability, any user of a web page would have the ability to download the fonts used on any given web page. This was a big problem for font designers and the foundries that represent them. Fonts represent valuable intellectual property, and distributing them freely through the web significantly devalues them.

From this need came a plethora of font-delivery systems. Font-delivery systems like Fontdeck, typekit, webType, TypeCloud, and Google Fonts, among many others, use proprietary code to deliver fonts to a user's browser without ever revealing the font files to the user. Now, designers can license and use fonts from a seemingly limitless library. And type designers and foundries can protect their intellectual property.

This makes the need for using web-safe system fonts less critical, but there are still very good reasons to rely on web-safe fonts: most web-safe fonts, like Georgia or Verdana, were designed specifically for screen use; they have design characteristics that make them more legible when viewed at small sizes on a screen; and the @font-face command, like imagery, can add load times to a page.

Branding.sva.edu/ uses the Google Font Enriqueta slab serif for the body text, a font that was created by combining robust and strong serifs from the Egyptian style with softer tones from Roman typefaces.

Fonts represent **valuable intellectual property** and @font-face leaves some question as to the end user's ability to reuse the font without paying for it.

Font styling is one of the most exciting and complicated areas of web typography. However, it is not the only area of focus for a designer. Web typography, like all forms of typographic expression, needs to illustrate a clear sense of hierarchy through the use of scale, color, and typeface. The examples shown here aren't meant only to dazzle with their typefaces, but to use the typefaces to convey a clear message.

KCCreepFest.com has a hand-drawn feel thanks to the Google Font Homemade Apple. The use of this font gives the site a feel of handwriting with the searchability of native type.

Grumpy wizards make toxic brew for the evil Queen and Jack.

Normal 400

Grumpy wizards make toxic brew

One morning, when Gregor Samsa woke from troubled dreams, he found himself transformed in his bed into a horrible vermin. He lay on his armour-like back, and if he lifted his head a little he could see his brown belly, slightly domed and divided by arches into stiff sections. The bedding was hardly able to cover it and seemed ready to slide off any moment. His many legs, pitifully thin compared with the size of the rest of him, waved about helplessly as he looked.

| FEATURE

100 Years Of Olympic Logos: A Depressing History Of Design Crimes

There's some beautiful graphic design on exhibit in these 45 Olympic Games logos, but most of them make you go WTF.

READ MORE >

41   1.1K   3.0K
NOTES   TWEET   LIKE

| FEATURE

100 Years Of Olympic Logos: A Depressing History Of Design Crimes

There's some beautiful graphic design on exhibit in these 45 Olympic Games logos, but most of them make you go WTF.

READ MORE >

41   1.1K   3.0K
NOTES   TWEET   LIKE

FastCoDesign.com uses a mix of MuseoSans, a sans serif text font, and FCZizouSlab, a custom display font for headlines. The combination creates a nice contrast between display and text type.

Co.Design
business +

A Designer Teach Pirates To Knock Off His Luxury

Editor: Sus

And in doing so, prove that you can never counterfeit quality.

READ MORE >

45   712   9.3K
NOTES   TWEET   LIKE

Comic Book Heroes Get A Gorgeous Native American Makeover

Batman, Superman, and Spider-Man look truly stunning following a traditional, Pacific Northwest makeover.

READ MORE >

10   566   15.3K
NOTES   TWEET   LIKE

Infographics Lie. Here's How To Spot The B.S.

Infographics are all over the place nowadays. How do you know which ones to trust? Follow these three easy steps to save yourself from getting duped.

READ MORE >

6   1.0K   2.0K
NOTES   TWEET   LIKE

Co.Design
business + innovation + design

Editor: Suzanne LaBarre

Subscribe to Newsletters

ENTER YOUR EMAIL:

submit

INFOGRAPHIC OF THE DAY

Can This Clever Statistical Model Predict Olympic Medal Winners?

EDITOR'S PICKS

100 Years Of Olympic Logos: A Depressing History Of Design Crimes

Subscribe to Newsletters

ENTER YOUR EMAIL:

submit

INFOGRAPHIC OF THE DAY

Can This Clever Statistical Model Predict Olympic Medal Winners?

EDITOR'S PICKS

100 Years Of Olympic Logos: A Depressing History Of Design Crimes

ENTER YOUR EMAIL:

submit

INFOGRAPHIC OF THE DAY

Between
ECONOMICS
AND LIFE
MARKETPLACE®
LISTEN NOW

Here again is the home page of HouseInd.com. The use of type, space, and color make it a stunning and inspirational example.

DESIGN & FONTS

Try Benguiat Caslon!

OBJECTS SHOP

HOUSE INDUSTRIES LETTERING MANUAL

THE PROCESS IS THE INSPIRATION

Fonts!

NEUHART DOLL PRINT

MUNICIPAL

BROWSE FONTS

WHAT'S UP AT HOUSE?

FIRST NAME   LAST NAME

ENTER YOUR EMAIL   ZIP CODE

I'M A   SUBMIT

STAY UP TO DATE ON NEW RELEASES, PRODUCTS AND EVENTS

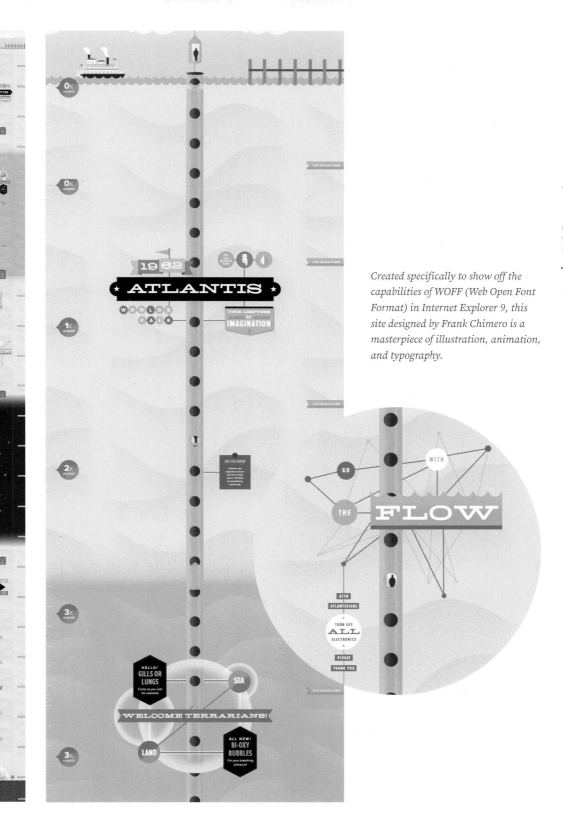

Created specifically to show off the capabilities of WOFF (Web Open Font Format) in Internet Explorer 9, this site designed by Frank Chimero is a masterpiece of illustration, animation, and typography.

The open-face design of the type on Fixate.it matches the line illustrations used throughout the site, creating unity between type and image.

IN CULP
Culpa qu

QUI OFFICIA DES
Officia deserunt m

DESERUNT MOLLIT ANIM
Mollift anim id est laborum. I.

ANIM ID EST LA
Id est laborum. La

EST LABORUM LOREM
Laborum. Lorem ipsum dol
ca sit amet, consectetuer adi

LOREM IPSUM M 10 M 10 SIT AM FI
Lipsum dolor sit amet, consectetuer ad
spiscing elit si co 0 s eiusmod tempor

Grumpy wizards make toxic brew for the evil Queen and Jack.

Grumpy wizards make toxic brew for th

Quo morning when simple barba woke from troubled dreams he found himself transformed in his bed into a horrible vermin he lay on his armour-like back and if he lifted his head a little he could see his brown belly slightly domed and divided by arches into stiff sections. The bedding was hardly able to cover it and seemed ready to slide off any moment. His many legs pitifully thin compared with the size of the rest of him waved about when he looked at.

line coming a piece

```
!  "  #  $  %  &  '  (  )  *  +
,  -  .  /  0  1  2  3  4  5  6  7
8  9  :  ;  <  =  >  ?  @  A  B  C
D  E  F  G  H  I  J  K  L  M  N  O
P  Q  R  S  T  U  V  W  X  Y  Z  [
\  ]  ^  _  `  a  b  c  d  e  f  g
h  i  j  k  l  m  n  o  p  q  r  s
t  u  v  w  x  y  z  {  |  }  ~  ¬
¢  £  ƒ  ¥  ¦  §  ¨  ©  «  ¬
·  ±  ²  ³  ´  µ  ¶  ·
»  ¼  ½  ¾  ¿  À  Á  Â  Ã  Ä  Å  Æ
Ç  È  É  Ê  Ë  Ì  Í  Î  Ï  Ð  Ñ  Ò
Ó  Ô  Õ  Ö  ×  Ø  Ù  Ú  Û  Ü  Ý  Þ
ß  à  á  â  ã  ä  æ  ç  è  é  ê
ë  ì  í  î  ï  ð  ñ  ò  ó  ô  õ  ö  ÷
÷  ø  ù  ú  û  ü  ý  þ  ÿ  –  —  ‘  '
'  "  "  „  †  ‡  •  ‹  ›  €
```

fixate
web & design

We are a web, illustration, & design agency

We create cutting-edge designs, websites, & web applications for companies & individuals who love the limelight.

We're a driven company with a simple goal – to make your business stand head and shoulders above the rest. We combine engaging design with state-of-the-art development, whilst always keeping the client's needs at the heart of every creative idea we have. We make it our duty to ensure the success of your business – after all – if you do well, a little of the light shines on us.

We are a web, illustration, & design agency

We create cutting-edge designs, websites & web applications for companies & individuals who love the limelight.

We're a driven company with a simple goal – to make your business stand head and shoulders above the rest. We combine engaging design with state-of-the-art development, whilst always keeping the client's needs at the heart of every creative idea we have. We make it our duty to ensure the success of your business – after all – if you do well, a little of the light shines on us.

set of work

In an increasingly digital world, a strong online presence is imperative.

We use our muscles to build intuitive and attractive websites that connect you with your customers. We specialise in thoughtful design that will enhance your brand and set you apart from the clamour of your competitors. All of our projects are executed at the highest standard, and we endeavour to remain ahead of the curve by keeping up to date with new and advanced technologies.

Publicity

We're thrilled at the positive response our site has received from people around the world! We've even been featured on some respected design and development sites and wanted to share a few with you.

Awwwards Site Of The Day – 7 Dec 2014
The Best Designs
Web Design Ledger
Zurb Responsive Website Gallery

© 2014 Fixate Web & Design | Web Design & Web Application Development by Fixate · Johan Weburg · South Africa

The designer of ChrisWilhiteDesign.com used bold condensed sans serif type contrasting with serif type to create a definite visual statement that is as distinctive as the products being shown on the site.

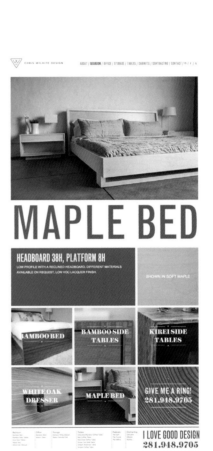

Display type on wineshop.hunters.co.nz is replaced using Cufón. Different than @font-face, JavaScript applications like Cufón use SVG graphics to display searchable type without using the font files.

Not all @font-face web typography needs to be big, bold, and in your face. This example, TheRivetPress.com, uses beautifully crafted details to create a distinctive and elegant look. The site uses Brandon Grotesque along with web-safe Georgia.

— presents —

THE RIVET PRESS

A HIUT DENIM PRODUCTION

ALL ARTICLES
View all The Rivet Press articles in one uninterrupted flow of consciousness.

SCRAPBOOK CHRONICLES
A carefully curated selection from the web that caught our eye and inspired us.

WORKSHOP WISDOM
Andrew Paynter visits maker's workshops to document the creative process.

HISTORY TAG TALES
Witness the lives and adventures of Hiut jeans through the eyes of their wearers.

THE FACTORY
News, media and all the latest behind the scenes from our small denim factory.

MAKERS & MAVERICKS
Our list of 100 movers and shakers that made a real change in 2013.

A COLLECTION OF WORDS
Essays, articles and articulate musings from Hiut Co-Founder, David Hieatt.

INSPIRATION
An all visual tumble of hand collected images and optical stimulation.

THE FACTORY

A Week In Instagram / No. 007

REFINE

01
If you run a company make sure it doesn't end up running you. You will have more ideas when you learn to switch off.

02
Sunshine trying to break on through.

03
Paul has brought in his old jeans to practice on. #freerepairsforlife #rawdenim

04
Storm blowing. Candles lit. Power cuts expected. #wildwest #cardiganbay

05
Calm. #notsowildwest

06
Good food plus good design equals good book.

Search our content for Artisans. We celebrate the skilled craftsmen and the makers.

Search our content for Denim. Whether Hiut Denim or others, it will all be here.

Search our content for Design. From architecture to stationary, everything has a designer.

Search our content for Farm. We farm ideas but we still get our hands & our Hiuts dirty.

Search our content for Food. A healthy mind requires a healthy appetite.

Search our content for Inspiration. Click the flame and light that fire.

Search our content for Music. Discover what tunes keep our Grandmasters sewing.

Search our content for Environment. Wherever you live, this is vital.

Search our content for Tech. The best and the brightest from the technological world.

SHARE

NEWER | HOME | OLDER

ABOUT
The Rivet Press is a Hiut Denim production, an online home for the Hiut community to grow, interact and get inspired.

In our factory we make jeans and we make them well. Online, we stitch ideas, unbutton dreams and press out the wrinkles. To read Hiut's story, click here.

MAKERS & MAVERICKS
At the end of 2013, the Hiut team got together to compile a list of the one hundred biggest influencers of the year. The people that made a difference, made a change, that went against the grain in favour of innovation and progress.

To see the full list, click here.

THE YEARBOOK

CONTACT
Email: stephanie@therivetpress.com

Instagram: @hiutdenim
Twitter: @hiutdenim
Google+: Follow
Facebook: HiutDenimCo

Sign up to the Newsletter

SITE BY THE PRINTER'S SON & POSITIVELY MELANCHOLY FOR HIUT DENIM

— presents —

THE RIVET PRESS

A HIUT DENIM PRODUCTION

ALL ARTICLES
View all The Rivet Press articles in one uninterrupted flow of consciousness.

SCRAPBOOK CHRONICLES
A carefully curated selection from the web that caught our eye and inspired us.

WORKSHOP WISDOM
Andrew Paynter visits maker's workshops to document the creative process.

HISTORY TAG TALES
Witness the lives and adventures of Hiut jeans through the eyes of their wearers.

THE FACTORY
News, media and all the latest behind the scenes from our small denim factory.

MAKERS & MAVERICKS
Our list of 100 movers and shakers that made a real change in 2013.

A COLLECTION OF WORDS
Essays, articles and articulate musings from Hiut Co-Founder, David Hieatt.

INSPIRATION
An all visual tumble of hand collected images and optical stimulation.

BORROWING
A passionate capture of the finest gear cycling culture in Bristol has been released for free online.

A WEEK IN INSTAGRAM NO. 007
Get up to date with the latest behind the scenes Hiut happenings from the cheery west coast of Wales.

WORKSHOP WISDOM
Hiut Yearbook photographer Andrew Paynter visits artists in their workshops to glean the processes behind the creative, celebrating the maker while simultaneously flying the flag for film photography.

Follow Andrew as he guides us around the private spaces of some of the most inspiring creatives, giving us a rare opportunity to see past the work to the artist behind it.

WILL GUITER / HIUT WEARER
We care about our jeans. We care about those where we design them, when we make them and even after we ship them. We want to know their lives and wearers. Meet Will, he's rocking a smallholding in his Hiuts.

INSPIRATION
Inspiration comes in all shapes and sizes, but no line sets sore and crawl. Every take on our ever evolving scrapbook of favourite images from around the globe and around the web.

ABOUT
The Rivet Press is a Hiut Denim production, an online home for the Hiut community to grow, interact and get inspired.

In our factory we make jeans and we make them well. Online, we stitch ideas, unbutton dreams and press out the wrinkles. To read Hiut's story, click here.

MAKERS & MAVERICKS
At the end of 2013, the Hiut team got together to compile a list of the one hundred biggest influencers of the year. The people that made a difference, made a change, that went against the grain in favour of innovation and progress.

To see the full list, click here.

THE YEARBOOK

CONTACT
Email: stephanie@therivetpress.com

Instagram: @hiutdenim
Twitter: @hiutdenim
Google+: Follow
Facebook: HiutDenimCo

Sign up to the Newsletter

SITE BY THE PRINTER'S SON & POSITIVELY MELANCHOLY FOR HIUT DENIM

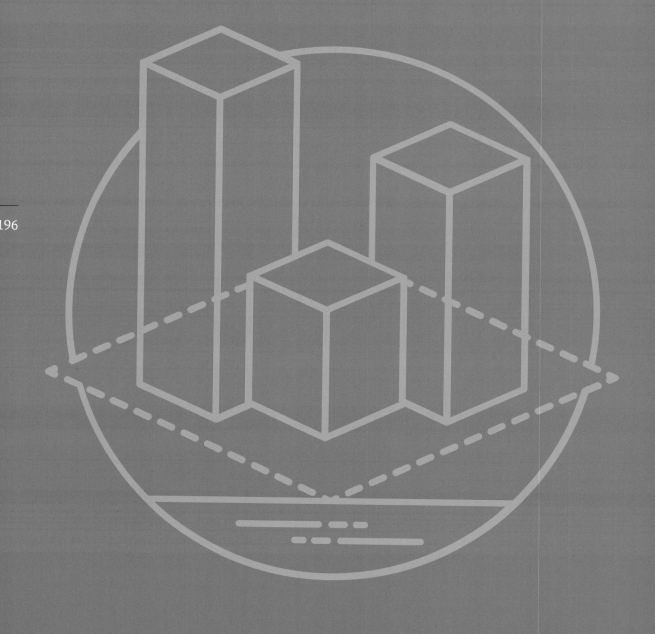

Section III

Optimize

Search Engine Optimization

Clients are always looking for the maximum financial return possible on their web project investment. Return on investment (ROI) is critical because developing a website can be quite expensive, and organizations need to show value for the money they invest in a web project. While design plays an enormous role in building a strong brand, and well-thought-out usability gives customers a great experience, neither matters if the target audience cannot find a site. Attracting the maximum possible number of site visitors is essential for the success of a site—and, in turn, the success of the company that owns the site. Simply put, getting found is everything to a business.

Getting Discovered: Browsing & Searching

There are three primary ways a user finds a specific site: by typing an address (URL) directly into the browser address bar; by browsing and following links or advertisements from one site to another; or by searching a topic in a search engine such as Google. While there's some debate over this topic, most research shows that well over half of internet users start by searching a topic using a search engine. This chapter explores the considerations one must make while planning, designing, coding, and promoting a site so search engines can find and index it.

Just like with web design and web usability, search engine optimization (SEO) is continually evolving based on trends and market factors. It would be difficult to codify specific techniques in a book whose usefulness is intended to last beyond the publication date. Therefore, this chapter focuses on the conceptual foundation of SEO—the basic principles that form the core of various trends. The exact techniques for a specific market or site can easily be found, ironically, by searching the web for SEO. Understanding why SEO is important, and the basic principles that influence effective results, helps a designer approach the planning and creation of a site with the correct mindset.

Types of Search Engines

There are two types of search engines:

Crawler based, like Bing.com and Google.com, which find sites using spiders to crawl the web and index content. A spider is a software tool that seeks out heavily trafficked servers for popular sites. Spiders are programmed to follow every link within a site while indexing the words it finds on each page. Crawler-based search engines gather information about a site and rank that site based on a series of on-site and external factors that will be explored in this chapter.

Directories, such as dmoz.org, rely on volunteer editors to evaluate sites for a specific topic and determine whether they should be listed in the directory. While these types of directories provide highly relevant sites, the process of selecting sites can be slow, which can result in some newer sites not appearing in the directory. Directory sites, however, can provide significant SEO value to a site that is listed with them. The inbound links (IBL) from popular directories to a site help to dramatically raise that site's ranking with crawler-based search engines.

For the sake of clarity in this chapter, it is important to define a couple of terms: a browser is an application installed on the user's computer and is used to browse and display web pages. Some popular browsers are Chrome, Safari, and Firefox. A search engine is a website or web utility that catalogs sites, through various means,

Top Crawler-Based Search Engines

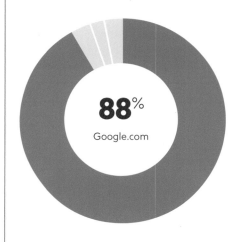

88% Google.com

Google	*88%*
Bing	*6%*
Yahoo!	*3%*
All other	*5%*

Source: Statista

Just like with web design and web usability, search engine optimization (SEO) techniques are **continually evolving** based on trends and market factors.

Google™

b bing

Google is by far the most dominant search engine, accounting for 88% of all internet searches. The next closest is Bing at 6%.

and presents the user with a list of sites that are relevant to the user's search. Some popular search engines are Google, Yahoo!, and Bing. A browser requires the user to know the exact domain name or URL of a site, beginning with www. and ending with .com, .org, etc., while a search engine requires only that a user have a topic he or she would like to find out more about.

The goal of any search engine is to sort through the millions of sites on the internet and deliver the most popular and relevant sites to a user based on a search term or phrase. The goal of a webmaster is to stand out from millions of sites and get his or her site listed however possible. It's an ever-evolving cat-and-mouse game where the rules change over time.

Early search engines relied mostly on site content when developing their rankings. A spider would simply read the text and the markup (the tags and code unseen by the user) to determine the content type and quality of a page. Some of these hidden bits of markup code include meta keywords, which can be listed in the <head> tag of the HTML and are intended to be the key terms and phrases used in the content of the page; the meta description, which is also found in the <head> tag and briefly describes the content of a page; and alt text (alternative text), which is a written description of a photo, for example, that can be translated to speech for vision-impaired users.

One issue with this method of cataloging is that these unseen tags can be filled with irrelevant terms that are nevertheless designed to yield high traffic. Say, for example, a webmaster launched a site for a brand-new widget. It's unlikely that anyone would be searching for this widget, so he might load the meta tags with terms related to cars—even though his widget has nothing to do with cars—because he knows millions of people search for terms related to cars every day. He may even put some white text talking about cars on his home page on a white background—white on white wouldn't be seen by the user but would be read as content by the spider. Within time, this widget site would begin appearing on searches for "cars"; however, when a user clicked on the link looking for car information, he or she would be disappointed to see that this site had no actual content related to cars—only widgets. This is called spamdexing or Black Hat SEO.

Additionally, as the internet evolved through the 1990s and into the 2000s, so did the types of content on websites. So-called "rich media," such as Flash, audio, and video content, cannot be indexed by search engines using typical methods. Since spiders cannot listen to, watch, or interact with content, sites that employ rich content were not ranking.

Search engines quickly caught on and began adjusting their methods of ranking sites to reduce spam, detect Black Hat tactics, and increase the ranking of sites that employed rich media like video. While some search engines continue to employ a site's meta description as the brief blurb under the link on a search results page, meta descriptions are not weighted heavily when ranking the site. Nor are meta keywords, alt tags, or other elements not seen by the end user, because of the ease with which they can be manipulated. Instead, search engines now use a combination of a site's popularity, in addition to its content, to determine their rank for the site. To do this, search engines not only look at on-site elements like title tags—the text that appears in the top of the browser window—but also off-site factors like the domain name's age and links from other sites. In fact, off-site factors have a greater effect on a site's rank than do on-site factors. By effectively understanding which sites are the most popular in relation to specific search terms, search engines can reasonably ensure that the content is relevant to users who search those terms.

Elements and Weight of Google Ranking

1. Link popularity

2. On-page keyword usage

3. Traffic and click-through rate

4. Anchor text of external links

5. Social graph metrics

6. Trust in the host domain

7. Registration and host data

Source: http://www.seomoz.org/article/search-ranking-factor.

The Wild West

As search engines get more sophisticated in their methods of evaluating sites, so too, do the individuals who intend to manipulate the results. A high search ranking can have a significant monetary value for an organization or an individual. As a result, the competition to reach the top with high-volume terms can get fierce. When money is involved, there's usually someone trying to cheat the system.

Honest, content-based methods of SEO, like those discussed in this chapter, are called White Hat techniques—named for the good guys who always wore white in the Wild West movies. Conversely, Black Hat SEO tactics are deceitful and manipulative. Those who practice Black Hat SEO are generally looking for traffic volume for its own sake—not to entertain, inform, or in any way provide value to the user.

Search engines retain the right to punish sites that practice Black Hat or deceitful tactics (knowingly or unknowingly) by removing them from their lists.

It's essential to select keywords based on the **customer's point of view**—not necessarily the client's internal vernacular.

Keywords

Before one can begin the process of implementing either on-site or external SEO techniques, one must first determine the best keywords for the site. Keywords are the specific terms that relate directly to the content of a site that people might use in their search. It's essential to select these terms based on the customer's point of view—not necessarily the client's point of view. Very often, clients speak of themselves using internal language—like product names or industry terms—that don't reflect how users search for information. It's important to understand how a user would define a client's business and use terms that fit that idea.

Users generally search for the solution to a problem they're experiencing: "What is [blank]?" "I need a [blank]," or "Where is [blank]?" Therefore, an effective strategy for developing a list of keywords is to position them as the answer to a question. These might be single words, but two-, three-, and even four-word phrases can be used. Identifying these words and phrases can involve a few methods.

Keyword tools such as the Google Keyword Tool, as well as third-party pay services, like WordTracker, help identify terms. These services are connected to a database of popular search terms and can cross-reference a specific term with other, synonymous terms that may have also been used to find sites related to the same topic. They can also provide information on the volume, popularity, and competition of terms as well.

Site-indexing tools crawl a site and provide a list of the current keyword mix. This is a good place to start implementing an SEO strategy on an existing site.

(Opposite) This is a screenshot from the Google Keyword Tool. Searching the phrase "web design books" produces the list of additional keyword ideas seen here. The list is helpful for determining the right balance of competition and monthly user searches—too much competition makes a word hard to target, yet too few monthly searches make a word less than valuable.

Keyword ideas — About this data

Keyword	Competition	Global Monthly Searches	Local Monthly Searches	Local Search Trends
web design books		9,900	4,400	
web page design templates		110,000	74,000	
web design templates		74,000	40,500	
web page design		165,000	110,000	
web page design tutorial		14,800	6,600	
web page design software		90,500	60,500	
flash web page design		368,000	165,000	
web page design tools		33,100	18,100	
web design courses		49,500	18,100	
learn web design		9,900	6,600	
professional web page design		18,100	12,100	
web design tools		14,800	8,100	
sample web page design		12,100	6,600	
web graphic design		60,500	49,500	
web page design jobs		18,100	14,800	
web page design layout		9,900	4,400	
freelance web designer		49,500	14,800	
learn web page design		4,400	3,600	
web page design ideas		2,900	1,900	
cool web page design		8,100	6,600	
web design studio		27,100	6,600	
web page design prices		8,600	4,400	
web design awards		18,100	9,900	
web design jobs		33,100	14,800	
web design company		201,000	90,500	
top web design		50,500	49,500	
flash web design		90,500	49,500	
web page design tips		3,800	1,600	
award winning web design		4,400	2,900	
web page design examples		12,100	6,600	
web page designer career		880	720	
custom web page design		22,200	14,800	
web design tutorial		27,100	9,900	
web design ideas		8,100	5,400	
personal web page design		49,500	40,500	
web design and development		90,500	40,500	
web page design cost		12,100	9,900	
web design software		165,000	74,000	
best web design		74,000	40,500	
good web design		12,100	6,600	
professional web design		60,500	33,100	
web design magazine		5,400	2,400	
great web design		5,400	2,900	
web design tips		12,100	5,400	
good web design examples		14,800	8,100	
artistic web design		720	590	
creative web design		22,200	8,100	
web design prices		27,100	14,800	
web design london		33,100	8,100	
web design services		110,000	60,500	

Sorted by Relevance — Views —

Go to page: 1 — Show rows: 50 — 1 - 50 of 800

Old-fashioned brainstorming, or role-playing—"If I were a user, what would I search?"—can produce a valuable list of terms that can act as a starting point before using a keyword tool.

When developing a list of key terms or phrases, it's important to think of broad enough terms, so there's an adequate amount of search volume, but not so broad that there are so many results that competing for the top spot would be impossible. For example, imagine a site that sells golf shirts patterned after retro shirts from the 1950s. Simply using the term *golf* would be problematic, since there are roughly 416 million search results for the term *golf*—everything from golf clinics and clubs to golf vacations and books. However, the phrase *1950s golf shirts* is too specific and may not yield the search volume that the client is looking for. Therefore, phrases like *classic golf shirts* or *buy retro shirts* might produce the right volume of qualified traffic with a reasonable ability to rank highly.

Keywords or phrases should not only accurately and specifically describe the content on a site; they should also be tailored to promoting conversion—a topic that's explored further in the next chapter. Most sites have a specific action they would like a user to take: sign up, buy, log in, etc. For these sites, it's not enough to simply be found: it's important to drive visitors who are looking to take action, so the keywords chosen for the site can include verbs like buy, as in the previous example, to promote high-value traffic—not just high volume.

Keyword lists should be kept at a manageable length—twenty-five to seventy-five words, depending on the size and type of site. A list that's too long can dilute the effectiveness of each individual keyword. Consistency and repetition is important for SEO, and a long list of words cuts down on the writer's ability to repeat terms. Although, it should be noted, some search engines may flag as spam a repetition of the exact same phrase numerous times, and this can be detrimental to a site's ranking. The terms that people use to search, and the concepts, ideas, and words used on a site, evolve constantly—and therefore so should the list of SEO keywords. The list should be revisited frequently enough to be sure all of the terms are current and connected to the user.

Designing for Spiders

Once a keyword list has been developed, it's time to begin employing those keywords on the site in ways that provide the most value for search engines. It's important to note that SEO factors shift in their overall importance, and no one factor will have a significant impact. It's the combination of these ideas and the management of them over time that creates an effective SEO strategy.

When designing for SEO, it's important to remember the two most basic things about how a search engine ranks pages:
- Is this page what it claims to be?
- How popular is this page?

The former is done by highlighting—visually and technically— specific key phrases that describe the page. The latter is done by linking to the page, as we will discuss later in this chapter.

While the majority of SEO techniques center around developing content and establishing relationships with like-minded sites, designers can have an impact on the SEO value of a site. Designing for SEO means using web-specific design methods, especially when it comes to displaying content, that yield visually interesting and dynamic results that search engines can index. This involves planning for an appropriate mix of graphics, animation, and content. Often, sites go too far toward one end of the spectrum or the other; too much of an SEO focus and a page can look generic or under-designed, while too much of a design focus, such as overuse of Flash or graphics for key text items, can result in poor search engine ranking. However, having an effectively optimized site doesn't mean it can't be designed well, and vice versa. It's simply a matter of employing the correct techniques.

The AIGA NY site employs many SEO best practices both seen and unseen. The site architecture is clear for easy crawling by spiders; links and headlines are filled with valuable keywords; and the source code is concisely written.

Designing for SEO means using **web-specific design methods**, especially when it comes to displaying content, that yield visually interesting and dynamic results that search engines can index.

In previous chapters, this book explored the pros and cons of various means of displaying type—or, more accurately, content. Using methods to display "live" text (as opposed to images of text) is important, but the concept of designing for SEO goes beyond just using web-safe type. The designer's arrangement of content is critical to effective SEO. Important, keyword-rich content should be displayed above the fold—the higher the better. The content should be broken up with headings and subheads, not only for scannability, but for SEO as well. Keyword-rich headings and subheadings should be styled using the "H" tags: <H1>, <H2>, <H3>, etc. The content in these tags is given greater weight by spiders since it is likely to contain information about the key ideas on the page.

Having keywords above the fold for the user to see is important, but equally—if not more—important is having keywords appear as high as possible in the HTML code for the spiders to find. To do this, pages should be built using CSS and <div> tags rather than tables. Using tables, an older method of building page structures with rows and columns, results in longer code that can push down content in the markup. The CSS styling should be imported from an exterior CSS style sheet to avoid having long stretches of CSS code in the <head> tag of a page. The same is true for JavaScript functions, or anything that can unnecessarily lengthen the markup.

Images can play a role in SEO as well. Since images saved by a designer are the exact same images that get downloaded by the user for display in a browser, the file names are important. Keyword-rich file names can help SEO—widget.jpg instead of img_123.jpg, for example.

Arranging content and creating assets in a way that's both user-friendly and spider-friendly is a unique challenge for a web designer. However, a designer can only have so much influence on the overall SEO strategy. An all-encompassing SEO strategy involves collaboration among a designer, a copywriter, the development team, the client, and even a media planner. What follows are other SEO factors that designers should be aware of, but which often are the responsibility of others on the team.

To the right are excerpts from the code for the web page seen on the far right, AIGANY.org. Only some of the important SEO features are displayed here, including:

• *Title tag, which appears as part of the browser window above. It contains valuable keywords that users might search to find a school.*

• *The meta description is used by Google and other search engines to describe a site.*

• *The <H1> tag is actually the logo on the page. The text is indented off the visible page and replaced with a background image of the logo.*

• *Subheads are styled as <H2> tags.*

• *Body copy is filled with linked keywords.*

```
<head>
  <title>Studio HMVD ~ Figuring it out - AIGA NY</title>

<meta name="description" content="Championing the future of design for all.">

</head>
<body>

<h1>Studio HMVD
Figuring it out</h1>

<h2>Bringing emotion to work, the physical to the digital, and carving out a niche in
a crowded scene.</h2>

<p>Building a studio from scratch is a choose-your-own adventure style endeavor.
<strong>Heather-Mariah Violet Dixon</strong> and <strong>Abigail Kerns</strong>
partners of women led <a href="http://studiohmvd.com/">Studio HMVD</a>
dive into the winding, textural process that has built their practice since 2018. From
exposing underlying fears at the start of projects, to their sense of play leading to
a whole new business niche, join us for a conversation on the highs and lows of
Studio HMVD.</p>

</body>
```

Internal SEO Factors

On-site SEO influencers can begin with the domain name or URL of the site. Finding the right domain name for a site can be difficult because so many names have already been taken, but choosing a completely arbitrary phrase could hurt a site's SEO value. Using a keyword in the URL can increase its relevance to certain topics. Also, the extension applied to the URL can affect its rank: .com and .org rank higher than other, less popular extensions like .me, .biz, or .us. The age of a domain can also play a role in a site's ranking. Similarly, keyword-rich page addresses can have a positive effect on SEO. For example, instead of naming a blog page using the date (www.example.com/2010/05/05/), the name should reflect the topic (www.example.com/widgets/widget_name/).

Developing a comprehensive SEO strategy means giving each page of a site an identity. This identity is formed and supported by key phrases or terms placed in strategic locations throughout the page. Not all key phrases will be used on every page—in fact, that's a common mistake. Instead, each page should focus on one or two key phrases to provide the most impact for the spiders looking to confirm that a page is what it claims to be. Spiders validate a page by weighing or giving more importance to certain elements over others, making the location of keywords critical to the SEO success of a page.

Probably the most significant location for a key phrase is in the <title> tag for the page. This is the line of text that appears at the top of a user's browser window, above the address bar. Crawl-based search engines place a very high value on this text, as it's very likely to reflect the content of a page. Therefore, the content of the <title> tag should be clear and to the point. Repetitive or non-descriptive <title> tags have a negative effect, such as simply repeating the name of the site on every page title.

TheBash.com is a site that allows users to book all types of entertainment. Their home page (right) contains text links to many categories of performers. These links combined with other internal SEO techniques consistently put TheBash.com at the top of search listings.

Ultimately, SEO is about content—valuable content.

Navigation plays a significant role in SEO. Terms that appear in links are given higher value by spiders. Therefore, it's important that the main navigation be styled using "live" text—as opposed to images—when possible. Breadcrumbing, as discussed in chapter 2, "Elements of Usability," is a great way to get keywords into links that appear on every page. Even the links within text play a part in SEO. When leading a user to another page, it's best to include keywords in the link ("Learn more about this widget" instead of just "Learn more"). Text-based site maps provide a useful tool for the user, but they also provide keyword-rich links for spiders. All of these keyword interlinks demonstrate to a search engine that the site, not just a single page, is rich with relevant content.

Ultimately, SEO is about content—valuable content. Each page of a site should contain at least some content; avoid landing pages or splash pages that simply lead a user to another page. (More on landing pages in a bit.) The content of each page should focus on a single key term or phrase and should be updated regularly. Syndicated content, or content that is being pulled from other sites via RSS (Real Simple Syndication), does not have significant SEO value; in fact, it can have a negative effect. Most importantly, content should be interesting to users. Users who value the content of a site generally tell others about it and even link to it from their sites or through social media. These links and high traffic can have a profound effect on the rank of a page.

External SEO Factors

As discussed earlier in this chapter, search engines have shifted their ranking methods away from focusing solely on site content to focusing on a site's popularity. The founders of Google developed this method by studying how college theses are evaluated. If a thesis paper is referenced by another thesis paper, it must have merit, and it must be truly about what the title says it is. The more theses that reference another thesis, the more valuable that thesis must be. Although the methods and details shift over time, that is how Google and similar sites rank web pages—by the number of sites that link to them. It's a form of validation.

Search engines look at the inbound links (IBL) that a site has—links that people use to connect to a site from other sites. The more inbound links, the more likely it is that a site is trusted. In addition to simply counting the IBLs, spiders read text within the links, and if it matches the content of the page, the ranking is boosted. The greatest value comes from two pages with similar content linking to each other. Similarly, but with a slightly lesser value, outbound links (OBL) to sites with relevant content can help with SEO. These links going out to other sites have lesser value because they can easily be manipulated by a webmaster. The goal, however, is to demonstrate that the site exists within a community of sites with connections back and forth.

The second way to determine a site's popularity is by evaluating the click-through rate, or the number of times a link has been clicked by a user, on a search engine results page (SERP). This is where the meta description tag for a site comes into play. While Google and other search engines no longer use the content of the metadata to rank a page, they do use the meta description as the blurb below the link on their results pages. A well-written meta description can help entice users to click.

(Opposite) This diagram illustrates the top six external and internal SEO factors. SEO factors shift and change over time, but the goal of a webmaster is to illustrate to a spider that a site is exactly what it's claiming to be.

A **comprehensive SEO strategy** targets high-value keywords with both an internal and external focus.

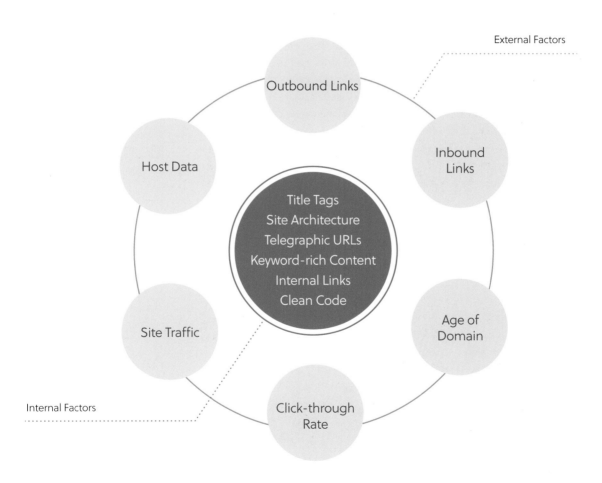

Finally, a site's popularity can be determined by its inclusion on directory sites. Since directories use knowledgeable human editors to evaluate the type and quality of content for a site, getting listed in a directory is a clear indication that a site lives up to its promise.

Paid Search

The concepts discussed to this point in this chapter produce what are known as organic search results—that is, the ranking of a site happened through the "natural" patterns and habits of users. There's another option to market and promote a site using search engines: paid search. It's called this because webmasters pay to have their sites listed on the search results page for specific terms. Paid search can be a valuable tool for marketers who are attempting to gain relevance in high-volume markets.

Paid search results appear at the top of most search results pages or on the right-hand side of the page. There is always some indication identifying paid search results, such as "Sponsored Links." This form of advertising can be sharply targeted to a specific segment of users, making it an attractive, relatively low-cost option for many clients. Pricing is usually based on the number of clicks an ad receives; this is why paid search is also called pay per click (PPC). Pricing is also based on the volume of the terms a campaign is targeting (the higher the search volume, the higher the price) as well as the position or "slot" that is desired—the top two slots cost more than the lower slots, for example.

The areas outlined in red are paid or sponsored search results. These links are paid for by advertisers targeting specific keywords—"SEO," in this case.

Paid search advertising can be **sharply targeted** to a specific segment of users, making it an attractive, relatively low-cost option for many clients.

Creating a paid search ad generally involves very little design—certainly none for the ad itself. Instead, ads are copy-based and styled by the advertiser. A paid search ad consists of a headline and body copy that conform to a strict word count. This word count and the need to fit keywords into the text make writing effective paid search ads a unique art form. Copy for an ad also has the ability to be dynamically generated based on a user's search term. For example, a PPC ad campaign for Florida might have a headline that reads, "Looking for vacations in [Keyword]?" When a user searches "Orlando Vacation Packages," the paid search ad would read "Looking for vacations in Orlando?" This gives the user the impression that the content behind this link is extremely relevant to his or her needs.

Paid search ads can link to a page within the advertiser's site, but for greater tracking and conversion, they can also lead to what's called a landing page. Landing pages are specifically designed to maximize the return on investment (ROI) for paid search and advertising campaigns. Often, two or three landing pages will be created to test which messages and design treatments work best. Over time, the pages with lower conversion rates are eliminated, again maximizing the ROI. The topic of converting browsers into buyers is explored further in the next chapter.

CHAPTER 8

Marketing and Conversion

This chapter explores various means of attracting users to a site beyond organic search traffic, converting them into valuable customers, and maintaining a profitable relationship with them. From paid advertising and viral marketing as means to attract visitors, to cross-selling and upselling and email marketing to keep them, each phase of the customer cycle can have a large and lasting effect on the number and value of users that come to a site.

Turning Browsers into Buyers

A website needs visitors in order for it to be seen as a success. Previous chapters have examined the methods of driving traffic through search engine optimization (SEO). SEO and search marketing sometimes aren't enough, especially when the client is looking to gain awareness among a specific target demographic for a product or service that's new or that fulfills a need that may not be obvious to a user. In these cases, a more proactive form of marketing is required—web marketing. Web marketing is a multi-billion-dollar industry covering a wide spectrum of services, from banner advertising and paid sponsorships to more organic forms of advertising like viral and social marketing.

When implemented properly, SEO, combined with effective web marketing, can drive large volumes of traffic to a site, but sheer numbers alone may not be good enough for a site to succeed long-term. Most sites require the user to take an action, from signing up to be a member, to buying things, to viewing as many pages as possible, to help with advertising impressions. Therefore, it's important that marketing efforts drive high-value visitors to a site. High-value visitors are visitors that come to a website not by chance or just to browse, but with the purpose of completing the required action of the site. Finding high-value users is a matter of promoting a site through the proper channels to target the right type of user, and by creating a compelling campaign that appeals to the needs of that target demographic.

Web marketing is a multi-billion-dollar industry covering a wide spectrum of services from banner advertising and paid sponsorships to more organic forms of advertising like viral and social marketing.

Browsers can be converted into high-value visitors, once they arrive at the site, through on-site marketing techniques. Certainly the methods of clear design and planned usability play a role in converting browsers into buyers, but there are other tools that a design team can use to further increase the conversion rate of users. Cross-selling is a means of telling a user, "If you like this, you might like that," and upselling is a means of telling a user, "This product is good, but that product will satisfy more of your needs." Both are effective ways to maximize the value of a user. Sharing mechanisms placed throughout the customer stream on a site can help spread the word about a site through word of mouth. This type of social sharing can be seen as significantly more trustworthy among potential users than banner advertising.

Once a customer has engaged with a client's brand by performing the required action on a site, the next step is to retain that customer. Retaining existing customers is vital for several reasons, but most important is the fact that it costs half as much to retain a customer as it does to attract a new one. Provided that an existing customer is happy with the experience, that person can help attract new customers by telling people about the experience and can even provide valuable feedback to the client about how to enhance the customer experience. Relationship marketing, which is used to communicate with existing customers, includes social marketing and email marketing. These elements help customers feel like they're on the inside and that they're appreciated.

Although entire books can and have been written on any one of these topics, this chapter gives an overview of the considerations a designer must make when attempting to add the most value for a client.

Display Advertising

Creating an effective banner ad campaign involves many disciplines, from copywriting and design to media strategy, technology, and even psychology. Users have become accustomed to tuning out banner ads, so getting noticed takes knowing the right techniques for a specific audience. As with any form of advertising, web banner advertising starts with the right media plan. A media plan is a strategy for determining where and when the banners will appear. These choices are made with several factors in mind, including the relevancy of the content on a site compared to the advertisement, the amount of traffic a site has, and the cost per click that a site offers.

Once in place, a media plan will dictate the types and sizes of interactive marketing units (IMU) needed for a campaign. The Interactive Advertising Bureau (IAB) has set standards for file size, dimension, and animation time. Included in the IAB standards are Universal Ad Package (UAP) sizes. UAP standards make it easier for companies to advertise, since advertisers only need to create a finite set of banner sizes that can be used across a wide range of sites. Universal Ad Package sizes (in pixels) include:

Leaderboard	728 x 90
Wide Skyscraper	160 x 600
Medium Rectangle	300 x 250
Rectangle	180 x 150
Mobile Leaderboard	320 x 50

A diagram showing the complete set of IAB IMUs is featured on the next spread.

IAB Ad Dimensions, File Sizes, and Animation Limits

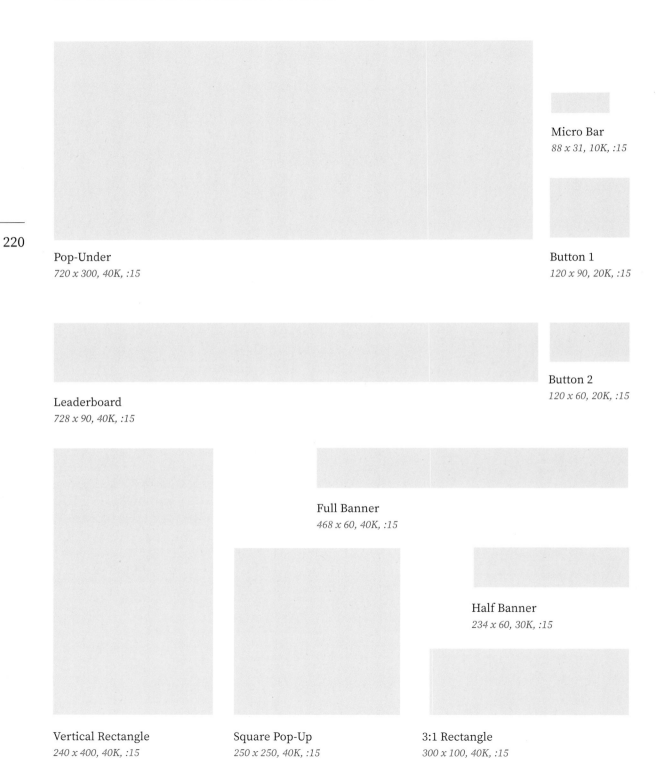

Micro Bar
88 x 31, 10K, :15

Button 1
120 x 90, 20K, :15

Button 2
120 x 60, 20K, :15

Pop-Under
720 x 300, 40K, :15

Leaderboard
728 x 90, 40K, :15

Full Banner
468 x 60, 40K, :15

Half Banner
234 x 60, 30K, :15

Vertical Rectangle
240 x 400, 40K, :15

Square Pop-Up
250 x 250, 40K, :15

3:1 Rectangle
300 x 100, 40K, :15

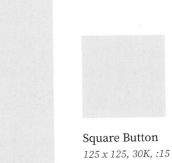

Vertical Banner
120 x 240, 30K, :15

Square Button
125 x 125, 30K, :15

Half-Page Ad
300 x 600, 40K, :15

Wide Skyscraper
160 x 600, 40K, :15

Skyscraper
120 x 600, 40K, :15

Large Rectangle
336 x 280, 40K, :15

Medium Rectangle
300 x 250, 40K, :15

Rectangle
180 x 150, 40K, :15

A **click-through rate** is the number of people who have clicked on the banner and is expressed as a percentage of the number of people who have seen the ad, called **impressions**.

When creating a banner ad, a designer is looking to generate a high click-through rate. A click-through rate is the number of people who have clicked on the banner to go to the client's site. The click-through rate is expressed as a percentage of the number of people who have seen the ad, called impressions. For example, if a banner is on a page where 25,000 people visit and 250 people click the banner, the click-through rate is 1 percent—an admirable rate for a site with this amount of traffic. This level of detailed statistical data is unique to web marketing, and it enables a high level of control over a campaign. Often, a banner campaign will involve multiple versions of a banner and over time, high-performing banners can replace low-performing banners to maximize the click-through rate of each placement.

Banners present a unique design challenge because they usually exist in a cluttered environment. These banners for the Starbucks Love campaign are instantly recognizable across different sites and the design is consistent throughout the varying UAP sizes.

Detach and Distribute

Because click-through rates are often a very small percentage of the overall impressions a banner receives, marketers have begun thinking about and utilizing the space within a banner differently. A technique called detach and distribute brings critical content and site features to the banner space, allowing users to engage with a brand without ever leaving the page they're on. Pioneered by Tom Beeby, creative director at the interactive marketing firm Beeby, Clark and Meyler, detach and distribute employs rich media to display a video, capture email addresses, or allow real-time social interactions, for example. This tactic of creating a mini-site within a site can be highly effective for increasing awareness of a product or service.

These banners created for GE display both pre-recorded and live video content from GE.com and allow users to comment on the them in real time, right within the banner space.

Contextually relevant ads are ads that respond directly to the environment in which they are served.

Context is a critical aspect of all forms of advertising, but with web advertising it can be taken to an even higher level. Contextually relevant ads are ads that respond directly to the environment in which they are served. This can mean something simple like placing an ad for fishing boats on a fishing website, but it can also be much more specific by drawing on data from the user, including time-specific or location-specific placements. Contextually relevant banners have been shown to be significantly more effective than one-size-fits-all banner campaigns.

Because of their unusual dimensions, shapes, file size limitations, and the need for immediate communication of a message, banner ads present a significant design challenge. The best advice a designer can heed is to put him- or herself in the shoes of the user and ask, "What would I respond to?" The answer is almost always a simple, relevant message, clearly stated, with an obvious call to action. Animation can help grab attention and/or build a message within a limited space, but most sites do not permit repeating or looping animation, since it can be very distracting to a user. Thus, the final frame of the banner should be designed and written in a way that all the critical information appears. The call to action, which is a sentence with a verb (learn, click, try, etc.) inviting the user to do something, should be clear—perhaps encased in a button-like object—and should directly relate to the content of the page the user is taken to after clicking the banner.

These banners from Apple Computer seem like ordinary ad placements, but there's a twist—the banners are synced with one another, making it possible for them to work together. In the ad seen here, Mac and PC are reacting to the leaderboard banner, which states that Apple is number one in customer experience, while the men in the seemingly unrelated "hair replacement" ad chime in to the conversation.

These amusing and engaging ads were awarded a Webby, one of the highest honors an online campaign can receive.

These ads for MySpace (top) and Pringles (left) use humor to engage the user and convey a brand message. This Pringles ad has received multiple accolades for its innovative use of adverting space. The ad continues seemingly forever with mundane conversation as part of Pringles' "Over-Sharing" campaign.

Getting a user to engage with a banner ad means getting a user to **engage with a client's brand.**

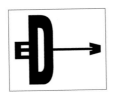

(Above) This single banner for Zippo lighters appears to be two banners, where the gentleman in the upper banner is being heated up by the lighter in the lower banner.

(Left) These banners for the Toyota Prius invite users to draw on the banner. This action triggers an animation that explains a feature of the car.

This interactive banner from Lotus Notes collaboration software invited users to collaborate by manipulating the letters of the word "IDEA" into various pictographs. Each user interacting with the ad would be responsible for shaping a single letter.

Rich media banners can be effective in grabbing a user's attention, but they can also be costly to produce and place, making them suitable for a limited number of clients.

In addition to standard ad units, there are third-party solutions, such as EyeWonder and EyeBlaster (Media Mind), which provide a variety of rich-media expandable banners. These banners include a wide range of interactive experiences, from a simple expanding banner, to banners that communicate with one another, to page takeovers or roadblocks where the entire web page is consumed with an ad. These banners can be effective in grabbing a user's attention, but they can also be costly to produce and place—so they're most suitable to a small number of clients who have large online advertising budgets.

These rich media banners for McDonald's completely take over the web page. The top image is known as a "peel-back" ad, where the page can be turned like a page of a book to reveal an advertising message. The bottom ad is an expandable banner featuring characters that dance across the screen.

(Top) This ad for Tostitos includes two standard placements, a leaderboard, and big box, and also the background "skin," which visually relates to the ads.

(Bottom left) This ad for Sony features an expandable video player. The player expands over the page content, making the video larger.

(Bottom right) This video game ad consists of a leaderboard that expands with a graphic and a video, as well as the big box ad along the right side.

Viral Marketing

Viral marketing gets its name from the way a virus spreads rapidly and "infects" a population organically. Viral marketing works because such pieces provide some sort of entertainment value beyond the thousands of ordinary advertising messages consumers are bombarded with on a daily basis. Successful viral pieces hit on a universal concept—humor, fear, sex—and at first may not appear to be marketing pieces at all. Branding is usually subtle, or in some cases nonexistent. Because consumers are so overloaded with advertising messages, they're also very suspicious, which makes viral marketing difficult—very difficult, in fact.

If a piece of marketing "goes viral," the impact can be profound. An early example of successful viral marketing was for the film *The Blair Witch Project.* Instead of standard big-budget TV and print ads, the producers released short clips of the film on the internet. The clips were hauntingly scary, and the supporting

website blurred the lines between what was real and what was part of the movie. The film cost $350,000 to create and market, but grossed nearly $250 million at the box office—the highest profit-to-cost ratio of any film in history.

The phrase "viral marketing" may be relatively new, but the concept isn't. Guerilla marketing, popular in the 1990s, involved tactics such as spray-painting company logos, as if by street artists, to get people talking and to gain credibility among an urban demographic. Even political propaganda or rumor-spreading can be considered a form of viral marketing.

Elf Yourself from OfficeMax allowed users to place family members' faces on dancing elves.

Have a break, have a Jesus Kit-Kat

Easter is time for Easter Bunny potato chips and Jesus sightings, and the latest is a doozy: Jesus has been spotted in a Kit-Kat.

The Kit-Kat hails from the Netherlands, where the story is a little Google Translate sketchy. Here's what I managed to pull out (original link/ translated)

Viral marketing doesn't have to be high-tech or high budget. This viral campaign from Kit Kat started with a photo and an email about seeing the face of Jesus. It quickly spread around the internet, carrying with it the Kit-Kat messaging.

Burger King and their interactive agency Crispin, Porter + Bogusky have a long history of creating viral content. Seen here is the subservient chicken who would do anything (really anything) the user typed into the field. Also seen here is The Simpsons Movie *tie-in*, Simpsonize Yourself. This Flash application allowed users to create Simpsons versions of themselves.

Brief History of Viral Videos

2000
John West Salmon

2001
BNW Films

2002
Agent Provocateur

2007
Charlie Bit My Finger

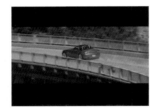

2009
David After Dentist

2008
Susan Boyle

2013
GoPro Camera

2014
First Kiss

2014
ALS Challenge

Successful viral marketing advertising hits on a universal concept—**humor, fear, sex**—and at first may not appear to be marketing at all.

2006
Dove Beauty

2006
Here It Goes Again

2007
GuitarMasterPro.net

2010
Old Spice

2011
Rebecca Black

2012
Gangnam Style

2015
Chewbacca Mask Lady

2017
Ping Pong Tricks

2020
Flatten the Curve

Social Marketing

Social marketing is similar to viral marketing in that it spreads organically through word of mouth—but social marketing usually involves a direct benefit to the user. Think of it this way: Viral marketing is a person going to a party with a cold and spreading it to the other partygoers; social marketing is a person going to a party with good news and actively telling as many people as he or she can.

Social marketing is used as much to get new customers as it is to retain existing customers. Building a social relationship with a customer by inviting them to like a page on Facebook, for example, enables client organizations to market to these consumers in a new way. Offering coupons or exclusive deals can make consumers feel as if they're part of a brand and therefore will be more likely to spread positive information about a brand to their social networks. These types of seemingly unaided endorsements have a profound ability to influence consumer opinion—so much so that companies are continually trying to blur the lines between "friends" and brands.

Social media isn't about fancy design; it's about engaging consumers on a different level than other forms of marketing. Social marketing is a conversation with the customer that makes the customer feel welcome and part of the client's company, as these examples illustrate.

Ben and Jerry's and JetBlue, whose Facebook and Twitter pages are seen here, respectively, do an excellent job extending their brand images with social media. This is in part because these brands already had a conversational relationship with their customers.

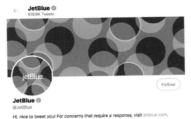

JetBlue ✓
618.9K Tweets

Search Twitter

JetBlue ✓
@JetBlue

Hi, nice to tweet you! For concerns that require a response, visit jetblue.com. Learn how we are keeping you safe and well: jetblue.com/safety

1-800-JETBLUE · jetblue.com · Joined May 2007

103.3K Following   1.8M Followers

Tweets   Tweets & replies   Media   Likes

You might like

Southwest Airlines
@SouthwestAir   Follow

americanair
@AmericanAir   Follow

United Airlines
@united   Follow

Show more

📌 Pinned Tweet

JetBlue ✓ @JetBlue · Jul 9
Why would you eat a "New York style" bagel anywhere else, when you could eat The Real Deal ...in New York?

Learn more: priceless.com/jetbluerealdeal

0:33 16K views

💬 212   🔁 28   ♡ 97

Show this thread

What's happening

US national news · LIVE
The $1.2 trillion bipartisan infrastructure bill passes the Senate
Trending with Senate, Infrastructure Week

Basketball · Trending
Dennis Schroder
6,492 Tweets

Rap · Trending
MAN WHAT
102K Tweets

COVID-19 · LIVE
COVID-19: News and updates for Connecticut

Soccer · LIVE
Lionel Messi reaches agreement on move to Paris Saint-Germain, French media outlets report
Trending with Neymar, Ramos

Show more

JetBlue ✓ @JetBlue · 4h
Book early and save big when you bundle flights + hotel by 8/12 for travel 9/1/21 – 5/31/22. ⚡ Min. $2,000 spend & terms apply. Book now: bit.ly/37gY/Hx

💬   🔁 4   ♡ 7

JetBlue ✓ @JetBlue · Aug 6
Our new nonstop service from JFK to London Heathrow takes off 8/11, and with England's new updated entry requirements, fully vaccinated U.S.-based travelers can travel without a quarantine (testing still required). Book now: bit.ly/2VzXveo

💬 37   🔁 15   ♡ 58

Pinterest.com has emerged as an engaging social media tool that companies are using to promote and grow their businesses. The visual nature of Pinterest allows users or potential customers to scan a lot of information very quickly.

Viral marketing is a person going to a party with a cold and spreading it to the other partygoers; **social marketing** is a person going to a party with good news and actively telling as many people as he or she can.

Perhaps the most famous and certainly one of the earliest social/viral campaigns was this one from Burger King. The Whopper Sacrifice called upon Facebook users to "sacrifice" a few of their friends for a free hamburger. The campaign was extremely successful; however, it violated a rule on Facebook that bans telling friends when they've been defriended. Because of this, the campaign was ended but its impact lives on.

Getting a user to take action involves the right products, promotion, pricing, and placement—**the four P's of marketing**.

On-Site Marketing

Once a user has found a site, it's important to the client that the value is maximized. Clients want to get the most out of each visitor, and this can mean different things for different sites—from becoming a member to filling a shopping cart with products to buy. Getting a user to take this action can take more than clear navigation, well-planned usability, and effective design; as discussed in previous chapters, it also involves the right products, promotion, pricing, and placement—the four P's of marketing.

Having the right product development and pricing strategy is largely the responsibility of the client and is usually determined prior to starting a web project. Promoting and placing these products, however, can be the job of the web project team. Promotion is a means of giving information about a product that piques the interest of the user. It's the job of an effective marketer to highlight important features of a product or service and clearly differentiate it from the competition. The web offers a variety of ways to promote a product or service, from photo galleries and slideshows to highly interactive product showcases.

The product display on MarieCatribs.com is not only user friendly but client friendly as well. The photography and clean layout make accessing the products easy and inviting, which can lead to more sales and higher profits for the client.

JaqkCellars.com does a magnificent job displaying their products in a way that enhances their appeal. The product pages are simple, with a single focal point: the product. Flash is used to provide a 360-degree spinning view of the bottle. The dark "ADD TO CART" buttons stand out from the page, making it easy for the user to enter the buying process.

To **cross-sell** is to recommend other products to a user based on his or her interest in a particular item.

The other P of on-site marketing is placement, which gives the user access to the product outside the context of the standard product or catalog page. Placement is the association of a product or service to content or other products or services. On a site that has health information and also sells health products, for example, an article about sprained ankles might be accompanied by a product placement of ankle braces for sale in the store.

Cross-selling is a form of placement. Online retailers understand that if a user is in the mood to buy one item, he or she is more easily persuaded to purchase more items. This is where cross-selling comes in. To cross-sell is to recommend other products to a user based on his or her interest in a particular item. Cross-selling associations can be done one of two ways: by the client linking products that relate to each other functionally—e.g., if you buy this Apple computer you might want this Apple mouse—or with purchase history, where users make the associations with their buying patterns—e.g., "Users who bought this item also bought..." Upselling is similar to cross-selling, except the goal is to get the customer to buy more expensive items or services. An effective way to upsell is through the use of a features chart. Features charts show side-by-side comparisons of one product to another, highlighting the benefits of purchasing the higher-priced item.

The items along the bottom this page from Glossier.com are related to the main product in some way and likely to be purchased at the same time.

PotteryBarn.com offers a variety of selling tools on their product pages—from items in a set and related items to customer ratings and reviews.

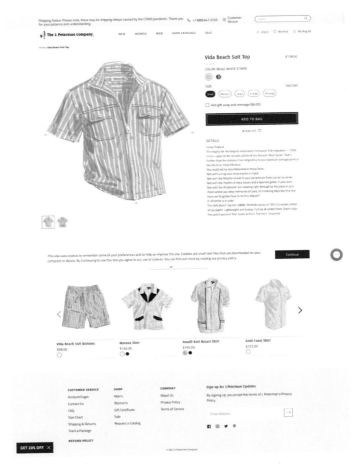

JPeterman.com cross-sell their garments by offering additional items of a similar style on the product pages.

Email Marketing

The site has been found and the sale made, but the customer cycle has one more component to close the loop: relationship building. Building a relationship with a user-turned-customer by regularly communicating with the person can be extremely valuable to a client. Repeat customers not only cost less than new customers, but they are likely to tell their friends about the product or service, which breeds new customers. One of the most effective ways to maintain a relationship with a customer is through email marketing. Email marketing "pushes" information about the client's product or service to the customer. Relationship marketing can take the form of a newsletter, where product information is accompanied by information that's valuable to the user.

There are strict laws governing the use of email marketing that designers and their clients should be aware of. Failure to comply with the laws contained in the CAN-SPAM Act can bring stiff fines to a client. The CAN-SPAM Act dictates the following guidelines for email marketing:

- Don't use false or misleading header information ("From," "Reply to")
- Don't use deceptive subject lines
- Identify the message as an ad
- Tell recipients where you're located
- Tell recipients how to opt out of receiving future emails
- Honor opt-out requests promptly
- Monitor what others are doing on your behalf

Source: http://www.business.ftc.gov/documents/bus61-can-spam-act-compliance-guide-business

TWELVE HAND-PAINTED PIECES
BY SARGIO SIGNS

FRAMED IN RECLAIMED BARNBOARD

PAINTED WITH ONE-SHOT LETTERING ENAMEL

VISIT THE 1151 GALLERY OR SHOP ONLINE

Adobe Creative Cloud

Tools to let you follow your intuition.

Your design workflow is about to get faster and more intuitive than ever. We've added new features in your favorite design apps that will expand what you can do while speeding up how quickly you can do it. Download the desktop apps today and see how the latest innovations in Creative Cloud can help you create your best work.

Photoshop CC

Download a free trial of the fastest, most responsive Photoshop yet and repurpose assets across Photoshop documents with linked Smart Objects. And experience Perspective Warp, the new feature that lets you change the viewpoint of a photo after it's been shot.

Get started

Illustrator CC

Finesse your designs more directly and intuitively with Live Corners. Round, invert, or chamfer one or multiple corners at the same time.

Get started

InDesign CC

Access the library of over 700 Typekit desktop fonts directly from the InDesign font menu (for complete membership only). The new missing font workflow even notes missing fonts in your documents, locates them in Typekit, and prompts you to sync them to your computer with a single click.

Get started

Adobe Muse CC

Create and publish dynamic websites for desktop and mobile devices — without writing code. Design using intuitive tools and add interactivity like scrolling effects, slide shows, contact forms, and videos.

Get started

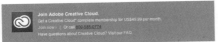

Join Adobe Creative Cloud.
Get a Creative Cloud complete membership for US$49.99 per month.
Join now > | Or call 800-585-0774
Have questions about Creative Cloud? Visit our FAQ.

Join the conversation:

These email templates from House Industries (left) and Adobe Systems (right) illustrate a simple elegance that's required for email designs. Because of the restrictions of mail browsers to display HTML and the need for immediate communication, email templates must focus on simplicity and clear hierarchy.

The **subject line** of an email acts as a headline and can play a pivotal role in the success of an email campaign.

Designing an email template presents another set of unique design challenges for web designers. This is because email clients (Outlook, Mac Mail, etc.) are far less sophisticated in their ability to display HTML than web browsers are. For example, the standard width of an email is 600 pixels, as opposed to 990 for a website. File sizes matter, since the user hasn't necessarily requested to see the content of the email. Emails with long load times tend to get deleted and go unread. Emails are primarily limited to HTML and standard image formats—jpg, gif—but Flash, JavaScript, and movie formats are currently unsupported by most email clients. Linking to external files for styling, for example, is also unsupported. Therefore CSS coding must be done "in-line," meaning in the individual tags for each HTML element.

The subject line of an email acts as a headline and can play a pivotal role in the success of an email. Subject lines should speak specifically to the subject of the email with clarity and brevity. Often, as with online banner advertising, multiple subject lines are tested for efficacy, and subject lines with higher open rates can replace more poorly performing lines to maximize the success of an email. Email layouts require simplicity even more so than web pages because they are often scanned by the user. When creating an email, a designer should consider the primary goal of the email and focus the design on that element by creating a clear hierarchy of information. Emails should include at least some HTML-based text because some email clients and mobile devices only display the text of an email. The footer of an email, by law, needs to indicate who the email was sent to, who it was sent by, and a means for the user to opt out from receiving future emails.

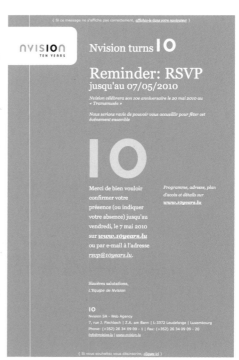

These email templates illustrate how, even within the constraints of email browsers, the design can still be an extension of a client's brand, increasing brand recognition among users.

Scott & Nix | Forward to a Friend

Take a Child Fishing

THE SUNFISH

Tips for a successful first outing with your young angler

1. Keep the trip short. **2.** Catch a fish.

3. Have a sandwich and a juice box on hand for when the fishing's done.

Pick a warm day, and take your youngster to a dock on a pond or a lake for sunfish. Beginnings are a delicate thing, and no one likes to get skunked, especially kids.

Trout and bass can be finicky and elusive. Sunfish are eager feeders and plentiful in most freshwater lakes.

Feel free to share our newsletter with your friends. Click here to forward it to someone you know.

The Quarry

It's nice to know what you're catching and sunfishes are an amazingly diverse group. All are members of the large Centrarchidae family, which includes freshwater basses, crappies, bluegill, pumpkinseed, and others. In all, there are 27 species and all are native to North America. The classic group of sunfishes, a.k.a. panfishes, are all included on the Sunfishes of North America wall poster.

SPECIES INCLUDE:

pumpkinseed
bluegill
spotted sunfish
orangespotted sunfish
redspotted sunfish
longear sunfish

redear sunfish
dollar sunfish
warmouth
redbreast sunfish
green sunfish
rock bass

flier
shadow bass
Ozark bass
black crappie
white crappie

SUNFISHES OF NORTH AMERICA

The Sunfish of North America poster is illustrated by the amazing Joe Tomelleri. You can see more of his illustrations on posters at our site, and at his website, americansunfishes.com.

The Gear

The Pole
Any light-weight fishing pole will do with a small reel and some 2- to 4-pound test line. You don't even really need a reel. This might be the time to use that old cane pole in the garage or to cut a branch from a willow and make one yourself.

The Hook and Knot
Use a number 6 hook, tied on the line using a clinch knot:

The Bobber
A simple one-inch adjustable bobber will do the trick. Place the bobber 18 inches above the hook.

The Bait
Earthworms (cut in small portions) are the traditional bait. You can also use live crickets (easy to catch in the cool of the morning), bits of soft pet food, small balls of white bread, mealworms, or even uncooked bacon. Sunnies will bite at just about anything.

Catching sunnies couldn't be easier and along the way, you can patiently explain some safety rules about hooks and how to gently release the fish back into the water. You will be rewarded with a very happy child and perhaps even the beginnings of a life-long angler.

The Technique

Toss the baited hook and bobber toward the shore or near the protective cover of the dock. Let it splash down and wait three seconds (counting it out with your child). Reel or pull the bobber back toward you, 12 inches, and let it sit. Keep an eye on it. It won't take long for the nibbling to begin. When the bobber goes under, give a slight tug to set the hook, and then slowly reel it in. Don't yank too hard, lest your child be unceremoniously introduced to a flying fish! If the bobber is just bobbing and not going under, try a smaller bit of bait.

Fish Stories

Before you get home, be sure to work out your story together. How big was the fish? How many did you catch? Fish stories are an integral part of the experience, and while we don't advocate fibbing, a little hyperbole won't hurt.

—Scott & Nix

Forward to a Friend   Subscribe to Our Newsletter   Contact Us

Scott & Nix | 150 W 28th St | New York, NY | 10001

CHAPTER 9

Analysis

The final component of the web design cycle is analysis. While all forms of marketing are analyzed and optimized, no form of marketing or design can be analyzed with the immediacy, accuracy, and depth that web marketing can. What used to take weeks or months to collect and report now happens in real time. This immediacy allows marketers and designers to make calculated adjustments that improve the overall performance of their online assets. From banner campaigns, to site design and usability, to email campaigns and social media, all aspects of user activity and brand engagement can be tracked at a granular level.

Closing the Loop

Website statistics have come a long way from the counters that used to be seen at the bottom of web pages. Those could only tell the webmaster the number of people who visited the site. Today, almost any action by a user can be tracked and analyzed—from where the visitor came; what words were used to search and find a site; how long the visitor was on a site; how many pages were visited—right down to if the person converted into a paying customer. Beyond the behavioral statistics, demographic information such as geographic location, browser type, OS, and connection speed can also be collected. Such statistics provide a marketing and design team with a wealth of useful information for optimizing site and campaign performance.

Analytical data can help remove a level of subjectivity from the creative process by providing qualitative data that supports one direction over another. Unfortunately, this data may not always support the designer's position. Web designers must be open to the notion that their designs will need to change and shift based on the habits and feedback of their users. What works for an audience today may be different next year, next month, or even next week. Technology evolves, users evolve, and environments evolve, making the web and web design more about progress and adaptability than permanence or even the level of perfection that comes with other forms of design.

The most common method of collecting statistical data is with Google Analytics, a free yet remarkably robust tracking system provided by Google.com. There are other free services, such as Matomo, which is a PHP-based open source system with many of the same features as Google Analytics. There are paid services, like WebTrends, that help their clients interpret their site statistics with reports and consulting.

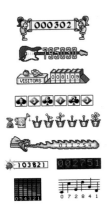

These counter icons are what webmasters once used to track users who came to a site.

In the case of web design, very often **progress** is more important than **perfection.**

This chapter examines various data points that Google Analytics reports on and how they can affect the decisions a designer makes. Each data point can provide valuable information, but the full potential of web analysis comes when the statistics are used in concert with one another. Focusing too heavily on any single statistic can mislead a designer. Combining key statistics can give a more complete picture of the strengths and weaknesses of a site. For example, if a site has a low average time-on-site statistic, this can be either a positive or a negative—but it's difficult to tell with this statistic alone. If the low average time on site is combined with a high bounce rate (the percentage of people who leave after only viewing the home page), then there could be an issue with engaging people in the site content. If, however, the low average time-on-site statistic is combined with a high number of pages viewed and the exit page leads a user to an online retailer to buy the product, for example, this would mean the site is working quite effectively in driving users to purchase.

What follows are brief explanations of various key statistics that Google Analytics reports on.

This is the Visitors Overview page of Google Analytics. By carefully tracking and cross-referencing the information displayed here, a designer can learn critical information about the habits of the users of a site and possibly inform future design decisions.

User Data

These data points tell a webmaster or designer what he or she needs to know about the users who visit a site. From the number of visits to the capabilities of the user's technology, understanding the user is critical to the success of a website project.

VISITS

This indicates the total number of visitors to a site. It includes new and returning visitors and is an indication of the success or failure of an SEO strategy or marketing campaign. The number of visits can be an overrated statistic in that it's not an indication of the value of the visitors in terms of how long they spent on the site or what percentage are returning because they liked the experience. Like most of the statistics in this chapter, the analyst needs to cross-reference the visits statistic with other statistics to really understand its value.

The term *visits* is sometimes confused with *hits*, but the two terms are not synonymous. A hit is a reference to the retrieval of a page asset from a server. For example, if a single user goes to a page with eight images and an external CSS file, each image plus the page and the CSS file will count as a hit—in this case, ten hits—but the page will have gotten only one visit. While hits have importance to an IT staff, designers and marketers should avoid citing hits as an indicator of a site's popularity, as it can represent a misleading and inflated view of site statistics.

ABSOLUTE UNIQUE VISITORS

Absolute unique visitors are visitors visiting a site for the very first time. Analytical reporting takes place over a specific time period. The default in Google Analytics is the past thirty days, but the range can be set for any length of time. Absolute unique visitors are not only visiting a site for the first time during the selected time period, they are visiting the site for the first time ever. This can be helpful in understanding the success of a marketing campaign whose goal is to build awareness among a new target audience.

NEW VISITS

New Versus Returning Visits is sometimes confused with absolute unique visitors, but there's a slight difference. New visits are visits to a site by users who have visited the site prior to the time range being analyzed, but it's their first time back during that time period. This data point is expressed as a percentage—56 percent indicates 56 percent of the visitors were new during the time period, and by inference, for example, 44 percent had visited the site more than once during the time period.

The image above shows the map overlay feature in Google Analytics. The darker the blue, the more visitors that have come to a site from that country.

BROWSER CAPABILITIES

The Browser Capabilities statistic shows both the number and percentage of browser types and technologies used by the visitors of a site. Understanding the capabilities of the majority of the users of a site is essential for designing and building the right experience for them. Included under browser capabilities is not only the browser type (Safari, Firefox, Microsoft Edge, Chrome, etc.) but also the operating system, screen resolution and colors, Flash version, and Java support. Each of these points paints a picture of the target users' capabilities and informs decisions made surrounding the types of technology used for a website.

NETWORK PROPERTIES

This feature indicates the service providers and hostnames of the users, but the most relevant data point for designers is the connection speed. Common connection speeds include (from fastest to slowest) T3, DSL, cable, ISDN, and dialup. Knowing the connection speed of the majority of the users of a site is critical to designing the right experience. The slower the connection speed, the lower the tolerance will likely be for graphics, imagery, and media assets that take time to download.

MOBILE

Increasingly, sites are being viewed using mobile devices, such as iPhones. This section of Google Analytics displays both the devices and the carriers of a site's mobile users. If a large number of visitors frequent a site via mobile devices, it may warrant a mobile version of the site.

MAP OVERLAY

Understanding the geographic location of the visitors to a site can play a role in informing the direction of a site. The Map Overlay feature of Google Analytics shows the countries where users have visited a site.

The intensity of the color indicates the number of visitors—the darker the blue, the more visitors. This allows web content developers to gear the content of a site in a way that is relevant to the users in the countries visiting the site.

LANGUAGES

Similar to the map overlay, the Languages report can help a client understand the needs of the actual demographic, which can be different from the target demographic. Languages are determined by the users' computer preferences and are reported in Google Analytics.

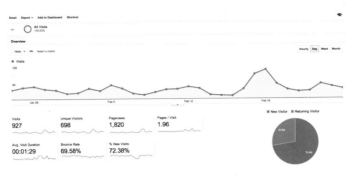

This visual shows the traffic source information. The pie chart indicates the three types of traffic sources: direct traffic, referring sites, and search engines.

Source Data

Once there's an understanding of the user, it's important to know how the user is finding a site. Source data plays a critical role in search engine optimization as well as marketing, because it gives a webmaster the knowledge of how users may have become aware of a site.

TRAFFIC SOURCES

Understanding the source of the visitor traffic to a site is critical for optimizing SEO and marketing efforts. Google Analytics' All Traffic Sources report shows the sources of traffic, including direct, search engines, and referring sites. The direct traffic number indicates users who simply typed the URL into their web browsers. This can indicate a number of things to a marketing team, including whether the user saw a web address in a non-online advertising campaign like print, radio, or TV.

REFERRING SITES

The Referring Sites tab shows sites that visitors used to link to the site being analyzed. This data is extremely valuable from an SEO perspective. The more sites that link to the subject site, the higher that site will rank for certain terms. Google Analytics displays the referring sites and the number of visitors that came from that site. By clicking on a site in the list, one can see the specific page the link came from.

SEARCH ENGINES

The Search Engine report shows the search engines that visitors used to search and find the subject site. This report can play an important role in determining the right sites for a search marketing campaign, as marketers want to advertise in the places where their target audience will see them.

KEYWORDS

The Keywords report is one of the most essential tools for understanding how users are finding a site. It shows a list of the words that visitors used to search for and link to a site. This can help validate or disprove an SEO key term strategy by showing the project team what words are actually being used to find a site. If the report matches the list of keywords the site targeted, the SEO strategy is a success. If they don't match, however, one of two things must occur. The team could look at the list and adjust it if there's an indication that the list misjudged what users were after. More likely, the implementation of the SEO tactics could be reexamined and improvements made to increase the performance of the original keyword list.

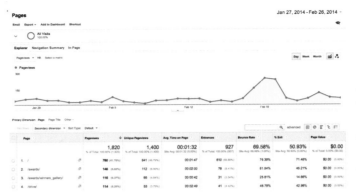

The content overview page on Google Analytics, pictured above, shows the pageviews, unique views, and bounce rates for a site.

Content Data

The final step in understanding analytical data is looking at what users are doing on a site. When combined, stats like landing pages, time on site, pageviews, and exit pages can give a clear view of how users are using a site.

PAGEVIEWS

Pageviews is as simple as the name implies—the number of pages viewed by visitors to a site. Pageviews is a broad statistic and, like total visits, can be somewhat misleading. For example, if a user reloads a page, that can count as a second pageview. Similarly, if a user browses from a page to another page, then back to the original, that too will count as two pageviews for the original page.

AVERAGE PAGEVIEWS

Average Pageviews is the result of the number of pageviews divided by the number of visits on a given day. This can be helpful in showing trends from day to day. Whereas Pageviews refers to the total number of pages viewed over the entire time period, Average Pageviews refers to the number of pages the average visitor viewed on a single day, which is then tracked over time.

BOUNCE RATE

The Bounce Rate is usually given as a percentage and indicates the percentage of people who left a site after visiting only a single landing page—the home page, for example. Generally, a high bounce rate is not a good thing. It can indicate that information is difficult to navigate, the traffic sources are misleading, or the content is of poor quality. In some rare cases, a high bounce rate is acceptable. For example, if a landing page effectively targets a specific keyword, a user may arrive at the page, get all the information needed, and then perhaps leave by clicking on a banner ad placed on the page. Despite going to only one page, that user might have a favorable opinion of the site and the client generated revenue with the ad click. More often, however, a high bounce rate is not good.

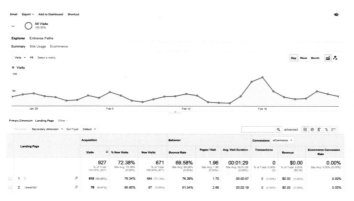

The Entrance Path *feature of Google Analytics shows where users entered a site and, based on that entry point, where they ended up.*

AVERAGE TIME ON SITE

Average Time On Site is, as the name implies, the average length of time users spend on a site. This statistic is calculated by subtracting the difference in time between the first and the last pageview. As a result, it can be somewhat inaccurate in terms of the exact amount of time users are spending on a site. If the last page of the visit involves a time-consuming task—which is usually the case if a user is watching a video or reading an article— then the time on site would actually be much longer. Designers and webmasters are looking for trends, rather than specific time, when analyzing the time-on-site statistics.

TOP LANDING PAGES

The Top Landing Pages are the pages that visitors are using to enter a site. Therefore, this data can be critical to a designer's decision-making process. It's important for designers to understand that not all visitors will be "landing" or arriving at the home page. With SEO and referring links, almost any page of the site can be a landing page. Designers need to provide the same type of marketing, usability, and accessibility on landing pages as they do on the home page.

TOP CONTENT

Top Content shows the pages on a site that were most viewed by visitors. This statistic shows the specific pages that were viewed and how many times they were viewed. This report also displays the average time users spent on each page, the bounce rate for each page, and the percentage of users who exited on a specific page. This can be helpful in gaining an understanding of what users want from a site. It can also help to show prospective advertisers where users are spending the most time when planning advertising sales.

TOP EXIT PAGES

Exit Pages are the last pages users viewed on a site. Users exit a site for various reasons—they've completed their task, or they clicked on an ad or link—or for less positive reasons, like they couldn't find what they were looking for or couldn't complete the required task. Together with landing pages and content statistics, exit page statistics complete the picture of how users arrive, what they do, and how they leave a site. Pages with unexpectedly high exit rates should be reexamined by the design team for usability issues that could cause users to leave the site prematurely.

In Closing

This is my third time writing and/or revising this book. I continue to be proud of the fact that the "principles" covered back in 2008 when I originally wrote Above the Fold are still relevant today. As technology advance and the landscape of the internet continues to change, the fundamentals of sound effective communication remain the same.

One area covered in this book that has seen the most significant changes is typography. Advancements in technology, specifically focused on preserving the rights of type designers, have made it possible to have an extremely wide selection of font options that wasn't at all possible thirteen years ago when I first wrote this book. But again, the fundamental principles of what makes a font effective for on-screen legibility and redibility has not changed, only the number of options available to the designer today.

With the proliferation of DIY design solutions and templates, the content of this book is even more relevant and valuable than ever. A good friend of mine, Michael Clark, said to me recently, "In the very near future, there will be exactly two types of workers: those who run machines; and those who are run by machines. And one pays significantly more than the other."

For web designers, this means if you've ever started a sentence in a job interview with "I can do [blank] in Adobe XD . . ." or "I'm an Figma expert . . ." the unfortunate, and somewhat counterintuitive reality is you fall in latter category of workers who are being run by machines. Limiting your skill set to the capabilities of a piece of technology means that anyone who watches a YouTube tutorial video or reads a how-to article can surpass you. It's the designers who proclaim, "I had a concept for [blank] . . ." or "the strategy behind this piece is [blank] . . ." that will excel now and in the future.

After decades of embracing technology as a means of executing design concepts, designers (smart designers) are learning that technology skills on their own have become a commodity. A commodity is something that has no differentiated value, like salt. My accounting professor used to say, "Salt is salt no matter who you buy it from, so buy it at the lowest price." Designers who are run by machines, and base their value on their mastery of a piece of software, are finding themselves in a race to the bottom of the pay scale—"I can do it cheaper than you so pick me." Not good!

Perhaps as a result of Mr. Clark's paradigm, there has been a widening schism in the field of graphic design over the past twenty years. On one side you have bargain suppliers of design commodities, on the other you have strategic business partners—and again, one pays more than the other. Strategic design partners understand the context of their work and the impact it will have on business results. Bargain suppliers can create a realistic drop shadow. Strategic design partners can articulate their role in the success of a campaign or a site launch. Commodity suppliers can articulate the difference between a PDF and a PNG.

Today, a student of graphic design will be forced (like it or not) to choose which side he or she will reside on. The concepts in this book will help move readers toward the former category of those who run machines. Knowing the "why" behind the "how" is what will separate those who excel from those who merely work.

Index

Books from Allworth Press

About Design by Gordon Salchow, foreword by Michael Bierut, afterword by Katherine McCoy (6⅛ × 6⅛, 208 pages, paperback, $19.99)

Advertising Design and Typography by Alex W. White, (8½ × 11, 224 pages, paperback, $29.99)

Citizen Designer (Second Edition) by Steven Heller and Véronique Vienne (6 × 9, 312 pages, paperback, $22.99)

Design Literacy by Steven Heller and Rick Poyner (6 × 9, 304 pages, paperback, $22.50)

Design Thinking by Thomas Lockwood (6 × 9, 304 pages, paperback, $24.95)

The Education of a Graphic Designer by Steven Heller (6 × 9, 380 pages, paperback, $19.99)

The Elements of Logo Design by Alex W. White, (8 × 10, 224 pages, paperback, $24.99)

Graphic Design Rants and Raves by Steven Heller (7 × 9, 200 pages, paperback, $19.99)

Green Graphic Design by Brian Dougherty with Celery Design Collaborative (6 × 9, 212 pages, paperback, $24.95)

How to Think Like a Great Graphic Designer by Debbie Millman (6 × 9, 248 pages, paperback, $24.95)

Looking Closer 5: Critical Writings on Graphic Design Edited by Michael Bierut, William Drenttel, and Steven Heller (6¾ × 9¾, 256 pages, paperback, $21.95)

How to Think Like a Great Graphic Designer by Debbie Millman (6 × 9, 248 pages, paperback, $24.95)

Emotional Branding: The New Paradigm for Connecting Brands to People, Updated and Revised Edition by Marc Gobé (6 × 9, 252 pages, paperback, $19.95)

The Art of Digital Branding by Ian Cocoran (6 × 9, 272 pages, paperback, $19.95)

Design Thinking: Integrating Innovation, Customer Experience, and Brand Value by Thomas Lockwood (6 × 9, 304 pages, paperback, $24.95)

Branding the Man: Why Men Are the Next Frontier in Fashion Retail by Bertrand Pellegrin (6 × 9, 224 pages, hardcover, $27.50)

Branding for Nonprofits: Developing Identity with Integrity by DK Holland (6 × 9, 208 pages, paperback, $19.95)

Brand Jam: Humanizing Brands Through Emotional Design by Marc Gobé (6 × 9, 240 pages, paperback, $24.95)

Citizen Brand: 10 Commandments for Transforming Brands in a Consumer Democracy by Marc Gobé (5½ × 8½, 256 pages, paperback, $24.95)

Building Design Strategy: Using Design to Achieve Key Business Objectives by Thomas Lockwood and Thomas Walton (6 × 9, 256 pages, $24.95)

Corporate Creativity: Developing an Innovative Organization edited by Thomas Lockwood (6 × 9, 256 pages, paperback, $24.95)

POP: How Graphic Design Shapes Popular Culture by Steven Heller (6 × 9, 288 pages, paperback, $24.95)

Designing Logos: The Process of Creating Symbols That Endure by Jack Gernsheimer (8½ × 10, 224 pages, paperback, $35.00)

Advertising Design and Typography by Alex W. White (8¾ × 11½, 224 pages, hardcover, $50.00)

The Elements of Graphic Design, Second Edition by Alex W. White (8 × 10, 224 pages, paperback, $29.95)

Designers Don't Read by Austin Howe (5½ × 8½, 208 pages, paperback, $19.95)

To see our complete catalog or to order online, please visit www.allworth.com.